Secrets of GREEN-SAND CASTING

"Lost Technology Series"
reprinted by Lindsay Publications

Secrets of
GREEN-SAND
MOLDING

reprinted from a volume published in 1906 by
International Correspondence Schools

ISBN 0-917914-08-2

9 0

INTRODUCTION

Technology gets more and more complicated in an effort to provide more and better goods at a lower cost. New technology provides impressive levels of productivity but requires very high investments. New technology is usually very expensive and very complicated — particularly for the home experimenter.

The backyard mechanic on the other hand is not interested in productivity. If in operating his backyard foundry, he makes one casting an afternoon and enjoys what he's doing, then he does not need the new high-priced technology. Old methods will do.

The Purpose of this book is to preserve old methods before they become lost forever.

So now you've built a cupola or small furnace and you have a ladle of molten iron or aluminum. Now what are you going to do with it? You can't very well pour it into a cake pan, bread box or an old boot. You've got to pour it into a sand mold where it will freeze into a useful casting.

You probably already realize that making green-sand molds is more of an art than a science. Where to put sprues and runners, vents for steam and gas, and just how hard to ram up the sand are skills that come only with practice.

Oldtimers will tell you that can't really learn green-sand molding from a book, and they're probably right. But this book comes as close to revealing the secrets as any ever published.

Use your imagination and the information here to greatly increase your skills as a foundryman, and keep alive some of the technology that is rapidly being lost to high-technology.

WARNING

Remember that the materials and methods described here are from another era. Workers were less safety conscious then, and some methods may be downright dangerous. Be careful! Use good solid judgement in your work. Lindsay Publications has not tested these methods and materials and does not endorse them. Our job is merely to pass along to you information from another era. Safety is your responsibility.

Write for a catalog of other unusual books available from:

Lindsay Publications
PO Box 12
Bradley IL 60915-0012

GREEN-SAND MOLDING.

(PART 1.)

IRON AND BRASS MOLDING.

INTRODUCTION.

1. Founding is a trade that involves some knowledge of almost every operation required in the making of machines; and men well versed in the mechanic arts assert that the art of founding demands greater mechanical skill, caution, and good judgment than any other of the allied trades. The art of founding is largely dependent on the hand, eye, and mind for results, machinery having played but a small part in the work of molders compared to what it has done for workers in most other trades.

2. There are three branches of molding, termed, respectively, *green-sand*, *dry-sand*, and *loam molding*. **Green-sand molding** involves the making of castings in molds that are composed entirely of sand in a damp state, or that have their surfaces " skin dried."

Dry-sand molding involves the making of castings in molds that are made with sand in a damp state, after which the sand is dried in an oven, or otherwise, so as to remove all moisture and leave the body of the mold dry and firm.

In **loam molding,** the castings are made in molds constructed with sweeps and skeletons of patterns; a mixture of loamy sand and other material is used to form the face of the mold, brickwork forming the outer and inner supports.

§ 40

This class of work, like dry-sand molding, requires thorough drying before pouring the metal into the molds.

3. The practice of some shops embraces all three branches, but most foundries make only green-sand molds. There is generally more risk in making medium and large castings in green-sand molds than in dry-sand or loam molds. In many cases a poor class of molders or inexperienced men may be employed for making dry-sand molds, but it is seldom wise to trust other than skilled workmen with the construction of green-sand molds, especially in heavy work. Loam work varies greatly in the degree of skill required. Some classes of loam molds permit the employment of inferior workmen, while others demand extraordinary experience, skill, and good judgment in their production.

DEFINITIONS.

4. The following are the definitions of some of the most common terms used in founding. They are given at this point so that they can all be found readily in case a student should care to refer to them and find the exact meaning of the terms when they occur later in the work.

A **flask** is a frame or box that keeps the sand in place while the casting is being made. Flasks may be made of wood or metal. A flask is composed of two or more parts. When composed of two parts, the one that stands underneath while the mold is being poured is called the **drag,** or **nowel,** while the portion that is molded last, and that stands uppermost while the casting is being poured, is called the **cope.** When a flask has more than two parts, the portions between the cope and the nowel are called **intermediate parts,** or **cheeks.** These terms are applied both to the parts of the flask and to the parts of the mold contained in the flask.

The **molding board,** sometimes called the **follow board,** is the board or plate on which the pattern is placed while ramming the sand into the drag.

The **bottom board** is the board or plate that is placed on top of the drag and fastened there before rolling it over, and hence it becomes the bottom of the mold during the subsequent molding and casting operations.

A **pattern** in connection with the foundry and machinery business is understood to be a form by the use of which a mold may be made.

A **sprue pin** is a wooden or metal pin used for making an opening in the cope through which the metal may be poured while casting. The opening formed by a sprue pin is called a **sprue** and connects the upper surface of the mold with a system of *gates* beneath. The sprue is sometimes called a **vertical runner.** The term is also applied to the body of metal that occupies the sprue passages after the casting is made.

A **gate** is an opening in the sand that connects the sprue, or **runner basin,** with the pattern. Gates may be horizontal or vertical, and in the case of some vertical gates may take the place of the sprue or runner basin. In fact, the terms gate and sprue are frequently used interchangeably. The term is also applied to the body of metal that occupies the gate passages after the casting is made.

A **draw-nail,** or **draw-spike,** is a sharp piece of metal used in drawing a pattern from the mold. It is driven into the pattern and holds by friction.

A **draw-hook** is a metal hook with a handle, used for drawing patterns that are supplied with special **drawing plates** to receive the end of the hook.

A **draw-screw** is a rod having a thread at one end, and is used for drawing patterns when they are provided with draw-plates attached to receive the draw-screw.

Drawing and **rapping plates** are metal plates fastened on patterns and intended to receive the ends of draw-hooks, draw-screws, or rapping irons. Frequently a draw-screw is used by screwing it into the hole in the draw-plate. The pattern may be loosened by rapping sidewise on the draw-screw, but it is better practice to use a rapping iron in a separate and unthreaded hole in the drawing plate.

A **riser,** or **feed-head,** is an opening from the cope face of a pattern to the face of the mold into which the surplus metal rises above the face of the casting.

Shake is the allowance that is made in the size of a pattern to permit its being rapped sidewise in order to loosen it in the mold so that it may be removed. Such an allowance is generally made only on patterns under 4 inches across.

Shrinkage is the allowance made on a pattern to compensate for the shrinkage of the metal in cooling.

Draft is a term used to denote that a pattern is tapered, the larger face of the pattern being at the parting line or surface of the mold. An allowance for draft is made to facilitate the withdrawing of the pattern from the sand, and results in the increase of certain dimensions of the pattern.

Molding sand is any sand used to make molds. It usually consists of a natural mixture of sharp sand and clay, the latter being necessary to make it adhere, or stick, together.

Parting sand is a general term applied to any material used to prevent two surfaces of a mold from adhering. It is usually sharp or burned sand.

Fireclay is any clay capable of withstanding intense heat. It is used for lining ladles, lining cupolas, and any place where great heat must be resisted.

Facing is a general term applied to any material used for lining the walls of a mold for the purpose of improving the surface of the casting.

A **rammer** is a tool used for tamping the sand in the mold. There are two classes of rammers—**hand rammers** and **floor rammers.** Hand rammers are used for light molding, and are only from 16 to 20 inches in length. Floor rammers are intended for heavy molding, and are usually several feet in length. One end of the rammer has a small rectangular point called the **peen,** and the other end a large flat surface called a **butt.** Rubber-tipped rammers are made, which are excellent for ramming close to the pattern and to insure even ramming.

A **sieve** is a tool used for sifting sand or for removing coarse material from the sand. The hand sieve is composed of a circular frame, the bottom of which is covered with wire cloth.

A **riddle** is a coarse sieve. The terms riddle and sieve are sometimes used interchangeably in the foundry.

A **trowel** is a flat metal tool provided with a suitable handle and is used in smoothing the sand in a mold or in place of a small shovel. There are a large number of forms of trowels used in molding that will be illustrated and described in connection with the work on which they are required.

Slickers are small trowels, or trowel-like tools, used for finishing the face of the mold, and will be illustrated in connection with the work on which they are required.

Gate cutters are small trowels or bent pieces of sheet metal employed for cutting the gates from the bottom of the sprue or down-gating to the opening left by the pattern.

A **vent** is any opening in the mold provided for the escape of gas or steam.

A **vent wire** is a small rod or wire used for forming the vent.

Rolling over is a term applied to the method of making a mold by making part of it in the drag, turning it over, and making the balance or upper portion of the mold in the cope.

Bedding in is a term applied to the making of a mold without rolling over, the first portion of the mold being made in the floor of the foundry. When all the casting is below the level of the floor, the cope is sometimes omitted, when it is said to be *open-sand casting*. In other cases, a portion of the casting may project above the floor, or a good surface may be required on the upper face of the casting, when it will be necessary to use a cope on top of the floor. In this case the floor mold is said to be *coped*.

A **core** is the name applied to any body of sand that projects into a mold. A core may be left by the pattern or may be formed separately and placed in the mold. In the latter case it is necessary to provide the pattern with core prints or projections to receive the ends of the core.

A **core box** is a box or frame in which sand is packed to form a core. The core box and the pattern are usually considered together as patterns for making a given casting.

MATERIALS.

MOLDING SAND.

5. Adhesive Qualities.—Molding sand is a mixture of sand or silica with a certain amount of clay or binding material that aids the sand in retaining any shape given it by pressure. The term molding sand is used to cover a large variety of sands employed in the different branches of green-sand molding. Sand is said to be *sharp* when its individual grains are angular, and *dull* when its individual grains are round. Sand is said to be *strong* when a body of it manifests a disposition to retain any shape that may be given it, and *weak* when it tends to fall apart and will not retain a given shape. Other things being equal, the sharper the sand, the stronger it is; but the sharpest sand is weak without some cementing material, as clay. Sharp sand alone will not hold its shape, while too strong a sand will not permit the gases to escape through it during the casting.

6. Grades of Sand.—The sand for making green-sand molds should vary in its physical qualities according as the castings to be made are light or heavy. For light castings the sand should be of a fine grain, while heavy castings require sands that are of an open, coarse-grained texture. If the sands that are best fitted for heavy work are used in making light work, the castings will have a rough skin, or surface. If the fine sands suitable for light castings are used for making heavy ones, there will be great danger of creating scabs or causing the castings to blow, on account of the sand being so fine in texture that it will not allow the gases (created at the face of the mold by heavy bodies of molten metal) to escape freely through the sand and vent holes of the flasks. There is also a tendency for fine sand to form a vitreous coating or scale on large castings, which can only be removed by pickling in acid.

7. Chemical and Physical Properties of Sand. Sand varies more in its physical properties than in its chemical composition. The chief constituent of sand is silica, though it contains alumina, magnesia, lime, iron, soda, and combined water. Different sands vary in the proportion in which these elements are combined. This affects the character of the sand for molding purposes to a large extent, which will be explained later. The variation in physical properties is more apparent and greater, but no more important. The sand may be fine or coarse, according to the size of the grains. The class or grade to which a certain sand belongs depends on this property in the sand. A detailed method for determining the fineness will be given later.

8. Silica.—Silica is the fire-resisting element; it has no bond, that is, it has no binding property; consequently, in a sand where adhesiveness is required, alumina must be present. Silica alone is very refractory, but in the presence of fluxing elements, it forms silicates that fuse or melt about as follows:

> Silicate of alumina melts at 4,350° F.
> Silicate of magnesia melts at 3,960° F.
> Silicate of lime melts at 3,810° F.
> Silicate of iron melts at 3,270° F.
> Silicate of soda melts at 1,500° F.

When soda or potash is present, silicates are formed at low temperatures. Iron melts at 2,200° to 2,300° F.; consequently, a sand containing much iron, lime, and alkali will burn or fuse into the molten metal. In other words, the more lime or alkali present, the more easily is the sand converted into slag.

9. Alumina.—Alumina causes the particles of sand to hold together; hence, a sand high in alumina is said to be strong, or possess bond. Alumina is very refractory, but, unlike silica, it bakes altogether like pottery at a high temperature; consequently, too much alumina must not be present in molding sand.

10.   Lime.—Lime may exist in sand as oxide, hydrate, carbonate, or sulphate, but usually as carbonate or oxide. The carbonate is the most objectionable.   Most of the lime salts are converted into oxide on burning; consequently, excess of lime will cause a mold either to drop or to crumble.

11.   Iron, Manganese, and Magnesia.—If combined iron is present, it may be converted into ferric oxide by heat and, in the presence of silica, produce slag.   Manganese in sand acts in a similar manner to iron, but is not so energetic.   Magnesia is very similar to lime, but less harmful on account of its being more refractory.

12.   Organic Matter.—Organic matter gives bond to sand, but the bond is destroyed by the burning of the organic matter as soon as it comes in contact with the molten metal, causing the sand to shrink and fall or crumble.

13.   Combined Water.—Combined water is always present in high alumina sands, and is one reason for the shrinkage in a strong bonded sand.

14.   Fineness.—Fineness of sand can be determined by the use of sieves, when a standard has been decided on. A good standard, and one that has been used to grade fineness, is to sift the sand through five sieves of 100, 80, 60, 40, and 20 mesh.   Exactly 100 grams of sand is sifted 1 minute in the 100-mesh sieve; the part that goes through is weighed, and the balance sifted in the 80-mesh sieve, and the process repeated on all the other sizes of sieves.   Any loss is credited to the 60-mesh sieve, and any that does not go through the 20-mesh sieve is credited to a 1-mesh sieve. The weights of sand going through each sieve are then multiplied by the mesh and the total divided by 100, which gives the degree of fineness.

The follöwing example will more clearly illustrate the method and calculations:

WEIGHT OF SAND PASSING THROUGH		NUMBER OF MESH OF SIEVE	
55.22 grams	by	100 mesh	5,522.00
20.89 grams	by	80 mesh	1,671.20
11.64 grams	by	60 mesh	698.40
10.57 grams	by	40 mesh	422.80
1.20 grams	by	20 mesh	24.00
.06 gram	by	1 mesh	.06
.42 loss	by	60 mesh	25.20
100.00 grams			8,363.66

Thus, 8,363.66 divided by 100 gives 83.64 per cent. as the percentage of fineness.

By this method the sand is graded into five grades according to its fineness.

	GRADE	DEGREE OF FINENESS
No. 1.	Superfine...........	Above 100 per cent.
No. 2.	Fine	90 to 100 per cent.
No. 3.	Medium...........	75 to 90 per cent.
No. 4.	Coarse or heavy.....	55 to 75 per cent.
No. 5.	Extra coarse........	30 to 55 per cent.

15. Quality of Sand.—Besides grading sand according to its degree of fineness, its quality should be determined for classification according to its chemical composition. A sand will then have a grade and a class that, taken together, will give a good idea of its value for foundry purposes. Three kinds of sand cover the field almost completely for classification as to quality. They are: *silica*, or *fire-sand*, *molding sand*, and *core sand*.

16. Silica, or Fire-Sand.—Silica, or fire-sand, is used for steel castings and where very high temperatures are necessary. Good fire-sand will usually run about 98 per cent. of silica, with very little alumina, lime, magnesia, or

combined water, and not more than a trace of iron.   The
following is an analysis of a good fire-sand:

Silica .... 98.04 per cent.
Alumina .... 1.40 per cent.
Iron.................... .06 per cent.
Lime................... .20 per cent.
Magnesia.............. .16 per cent.
Combined water .... .14 per cent.

Total............... 100.00 per cent.

Specific gravity, 2.592.

17. Molding Sand.—Molding sand for ironwork generally contains from 75 to 85 per cent. of silica, 5 to 13 per
cent. of alumina, usually less than 2.5 per cent. of lime and
magnesia, not over .75 per cent. of fixed alkali, generally less
than 5 per cent. of iron, and seldom more than 4 per cent. of
combined water.

18. Core Sand.—The quality or chemical composition
of a core sand according to some authorities is of minor
importance, the degree of fineness being the main feature.
As a rule, a good core sand should be high in silica and low
in alumina.   The bond for core sand is obtained by adding
resin, flour, etc.; consequently, the desired effect is produced with a high silica sand or with sand low in alumina
and iron.   A sand low in alumina and iron will permit the
gases to escape rapidly, whereas a high alumina or a clay
sand bakes and holds back the gases.

19. Localities in Which Molding Sand Is Found.
Molding sand, suitable for medium-weight and heavy green-
sand castings, is found in almost every part of the United
States.   Sand for light work is the most difficult to obtain;
for many years light-work foundries were compelled to rely
wholly on the fine sand found in and around Albany, New
York, but sand suitable for such work is now found in many
states.   Sand for statuary work in bronze is still imported
from France, no suitable substitute having been discovered.

TEMPERING THE SAND.

20. Meaning of Term.—In mixing or **tempering**
sand by hand, the
shovel should, be used
in such a manner as
to scatter the sand,
as shown in Figs. 1
and 2. This is done
by giving the shovel
a twist with the hand
that holds the handle
end. When shoveling
sand from one place
to another without at-
tempting to mix it,
the sand is sometimes
allowed to leave the
shovel in a solid mass,

FIG. 1.

as shown in Fig. 3. This method of shoveling permits the

FIG. 2.

sand to be thrown to a greater distance, and hence is used
when shoveling sand from place to place, as from a car to a bin.

A molder or helper should learn to shovel either right- or

FIG. 3.

left-handed, so as to be able to take either side of a sand
heap when working with an assistant; note the different

FIG. 4.

relative positions of the two men in Figs. 2 and 4.   A clear space of 1 or 2 feet should be maintained between the pile from which the sand is being shoveled and that on which it is thrown, as shown in Figs. 3 and 4. If this is not done, much of the sand will escape thorough mixing.

21. Shovels.—In Fig. 5 are shown the two kinds of shovel in general use, (*a*) having a flat blade, while (*b*) has

FIG. 5.

turned edges. The flat shovel is generally used for light floor work and bench molding, while shovel (*b*) is used for heavy molding and digging out holes.

A molder should never work with a dirty shovel. When in use the shovel should be kept clean by scraping as shown in Fig. 6. When put away for the night, or if not in constant use, a shovel should be cleaned of all dirt and oiled with a greasy rag to prevent its getting rusty. There is nothing that denotes the poor or slovenly molder more than working with a dirty shovel.

FIG. 6.

22. Wetting Down the Sand.—In throwing water on a sand pile with a bucket or hose, it should never be thrown in a body on one spot, as that will form mud-holes and involve a loss of time and labor in mixing the mud with drier sand in order to temper it. If the sand is very dry, water should be sprinkled upon it from a hose, or by being thrown from a bucket, as

FIG 7.

FIG. 8.

in Fig. 7. When the sand has been dampened nearly to its right temper, so as to require but little more wetting, it should be sprinkled by hand from the bucket, as shown in Fig. 8.

23.   Sieves and Riddles.—The meshes or openings of a **sieve** range from the fineness of a flour sieve up to openings $\frac{1}{8}$ inch square; when the openings are above $\frac{1}{8}$ inch square, the implement is called a **riddle.** In Fig. 9, (*a*) shows a sieve and (*b*)

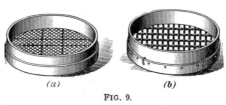

(*a*)   (*b*)

FIG. 9.

a riddle. In sieving or riddling sand by hand, the riddle should not be held rigidly, as shown in Fig. 10, but should

FIG. 10.

be held loosely, so that the butt of the hand can strike, or jar, the rim of the riddle as it is swung from one side to the other, after the manner shown in Fig. 11. Hitting the rim of the riddle or sieve with the butt of the hand causes a jar

over the whole face of the meshes, and this causes the sand
to pass through the meshes much more freely than if it were
held rigidly, as in Fig. 10.

FIG. 11.

When sieves or riddles are not in use, they should not be
thrown on the damp ground or on a sand pile with the mesh

FIG. 12.

side down, as such a
practice will clog the
meshes with sand in such
a manner as to hinder
the passage of sand
through the screen and
also cause the wires to
rust away very rapidly;
they should be placed
either on the sand heap
with the mesh side up,
or else hung upon a nail,
as shown in Fig. 12. A
skilled molder can be de-
tected by the way in
which he handles his
sieve and riddle, as well as his shovel.

PREPARING THE MOLD BY ROLLING OVER.

24. Distinction Between Methods.— The term *rolling over* comes from the fact that one of the processes of making a certain class of molds consists in turning the lower part of the mold completely over, so that what was the bottom is the top. As used in foundry practice, it refers to all molding in which the lower part of the mold is turned over, and distinguishes this class of work from that in which the lower part of the mold is made in the foundry floor without a flask. The molds so made cannot be rolled over, but as they are rammed in the sand of the floor, the term *bedding in* is used and seems especially applicable. Molding by rolling over will be considered in the following pages, while bedding in will be discussed in following sections.

25. Flasks are open frames of wood or iron, usually consisting of two parts, though sometimes of three or more, so arranged that they may be taken apart and then returned to exactly the same relative positions that they had before they were separated. For this purpose they are provided with special projecting pins and sockets. There is at least one molding board for each style of flask and as many bottom boards as there are flasks. When there are two frames (the cope and the drag) to the flask, it is called a "two-part flask"; when three frames, a "three-part flask"; and so on, for four or more. There is a great variety in the shapes and sizes of flasks; the molder usually takes the smallest flask in which he can safely make the mold for any given casting. There is also a special type, known as a "snap flask"; this opens at one corner and has a hinge on the opposite corner, so that the flask may be removed from a mold, after the mold is completed, and used over again instead of remaining until the metal is poured into the mold.

RAMMING THE MOLD.

26. Use of Rammer.— Fig. 13 (*a*) represents the ramming up of a deep pattern *P*. The mold is rammed up in courses, and the depth of the ramming courses varies from

4 to 8 inches, according to the strength of the molder and the character of the work he has in hand. For the pattern shown, a depth of 5 to 6 inches of loose sand will work well in the ordinary molder's hands.

In starting on a mold, the patterns should be faced with a proper **facing sand** that has passed through a No. 8 sieve, so as to have an even surface on the face of the mold. The facing sand is applied to the face of the pattern for a thickness of from 1 to 1½ inches, as shown at *h*, *h*, Fig. 13. After

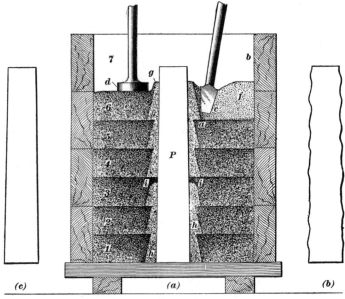

FIG. 13.

this facing sand has been banked against the face of the pattern, common heap sand is filled in to the level of the facing sand, and the whole is then rammed down with the peen and butt of the rammer. In banking facing sand against a pattern, it should be carried as high as the common sand is intended to go. If this is not done, the result will be a seam of common sand, as at *i*, Fig. 13, separating the parts of the facing sand at the surface of the pattern *P*. If the facing sand is not thoroughly united at the seams of each course, the castings will be rough and will often have "burned sand swells" at the seams of all such courses. This will make a casting look almost as badly as do swells caused by uneven ramming, as shown in Fig. 13 (*b*).

In learning how to use the **floor rammer,** beginners generally handle it by placing one hand on top, as shown in Fig. 14, which is wrong; the proper way to handle it is

FIG. 14.   FIG. 15.

illustrated in Fig. 15.   In starting to ram a course of sand, the peen end is used first and should be struck down with sufficient force to compress the sand until it feels solid, showing that it is down to a bottom.   This bottom may be either the mold board *c*, Fig. 15, or it may be the course of sand that has just been rammed, as *5* in Fig. 13, where the peen end of the rammer is shown at *c* and the rammed sand of the previous course at *a*.   In ramming the space between the pattern and the inside of the flask *b*, Fig. 13, the peen is first carried along the face of the pattern, as shown at *c* in the illustration, and then used promiscuously over the remainder of the space until it has been rammed down fairly well with the peen; then the rammer is reversed and the butt *d* used on the surface.

After the butt has been used to ram down a course of sand 5 or 6 inches thick, it leaves that course from 1 inch to

1½ inches lower than when finished by the peen, as will be seen by comparing the portions *e* and *f* of course *6*, Fig. 13. The mound next the pattern seen at *g*, Figs. 13 and 16, is then rammed down lightly with the peen. After this, the hand is used, as shown on the right of Fig. 16, to level down and clean away the facing sand lying against the pattern (which is liable to have common heap sand mixed with it), to prepare for another course of facing and ramming. In using the rammer for the first

FIG. 16.

peening and the butting, care should be exercised to keep the peen and butt about 1 inch from the face of the pattern during the ramming. When driving the peen down, it is liable, if forced up against the pattern, to make a hard spot in the face of the mold that may cause a scab on the face of the casting at that point. If the courses of sand are not rammed evenly, after the manner just described, the castings are very liable to be badly swelled at the line of union of the courses, and thus appear as shown at (*b*), Fig. 13; whereas, if the courses of sand are rammed evenly, the casting should appear as shown at (*c*).

27. Hardness Required in Molds.—As a rule, the sand on the sides of molds will stand harder ramming than that on the bottom or the top; and again, the deeper the mold, the harder the ramming required at the lower courses. As an example, in ramming the mold shown in Fig. 13 (which is rolled over when rammed full), the courses *5*, *6*, and *7* will require harder ramming than the courses *1*, *2*, *3*, and *4*, because the pressure of the metal is greater near the bottom. Molten metal behaves like water in this respect; if we fill a barrel with water, there is greater pressure on the sides near the bottom than near the top. It is the same with molds; the deeper they are, the harder should the sand be rammed and its support strengthened to prevent the static pressure of the metal bursting the bottom of the mold and letting the metal run out. The extra degree of hardness at the bottom portion of the mold, as compared with that near the top, depends on the power and time applied in ramming with the butt of the rammer. The face of the mold should not differ very much in hardness at any point in the height of the mold. This evenness is obtained by treating the mold with the peen in the same manner at the bottom as at the top. If the mold is very high, the lower portion should be rammed a little the harder. By keeping the edge of the butt of the rammer back about 1 inch from the face of the pattern, as already explained, some hard butt ramming can be given to the common sand back of the facing sand without getting the face of the bottom portion of the mold so hard that the metal will not lie against it. By butt ramming all the surface, up to a point about 1 inch from the face of the pattern, until it is hard, it is possible to make the entire face of the mold harder than if lighter butt ramming were done close to the face of the pattern. Nevertheless, this plan will not make the face of the mold as hard as if greater force were also applied to the peen when ramming. Castings having a depth of 4 or 5 feet have been made in green-sand molds; at these depths the back support of the molds must be very rigid and the ramming nearly as solid as it can be made with the butt. Were the upper portions of such molds rammed as hard as the bottom must be, the castings would blow and scab.

The reason why the lower portion of a mold will permit so much harder ramming than the top portion is that the greater pressure of the metal at the lower portion forces the steam and gases to escape through the pores and venting of the sand. At the top of the mold there is not sufficient pressure to do this. If a mold is not sufficiently soft at its upper end to allow the steam and gases formed at the face of the casting to escape freely through the facing sand to the rear sand and its vents, they will pass out through the metal. When steam and gases escape in this manner, they are liable to cause a mold to blow. This may result in scabbing a casting or in spoiling it. It should always be remembered that steam and gases will pass off in the direction in which they meet with the least resistance.

DEEP MOLDS.

28. Sand Required.—While it is true that the lower part of deep molds requires hard ramming to prevent the molds from bursting or being changed materially from the size of the pattern, there are conditions other than that involved in the ramming that must be provided for and controlled. In using facing sand, as well as in using common heap sand, in ramming deep molds, the sand must be worked as dry as possible and must be freely vented. If the damp sand that can be safely used on the sides of shallow molds is used for deep molds, and rammed as hard as it should be, it becomes a source of danger. The chances are that the moment the metal flows into the deep mold, it will commence to blow and may throw out the metal with considerable force, spoiling the casting and possibly burning the workmen standing near by. The sides of a mold may be rammed much harder if the sand is comparatively dry than if it is wet. The reason for this will be explained further on.

29. Venting the Mold.—The sides of all deep green-sand molds must be well vented. The harder the ramming, or the damper the sand, the greater is the amount of venting required. There are several methods used in providing these necessary vents. Suppose the pattern in Fig. 17 to be bedded in instead of being rolled over, as in Fig. 13. In

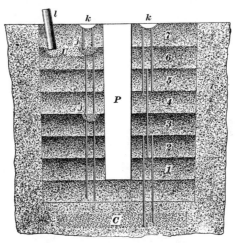

FIG. 17.

ramming the mold, the same plan is followed as in Fig. 13, the only difference being that the courses *1*, *2*, and *3* are

rammed the hardest, being the ones that must stand the greatest strain; whereas, the courses *1*, *2*, and *3*, in Fig. 13, being the uppermost when cast, are rammed the lightest. In ramming up the pattern, Fig. 17, by bedding in, the sides can be vented in the following manner: After three to four courses have been rammed, a channel or groove is cut out of the solid rammed sand, as shown by the letter *j* in the course marked *3*, Figs. 17 and 18. This being done, the channel

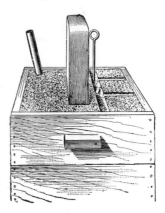

FIG. 18.

is vented with two rows of $\frac{1}{4}$-inch vents. These vents are put in 2 inches apart and are kept back about 2 inches from the face of the pattern; their position is shown by the rods seen projecting from the sand at *j*, Fig. 19, in which illustration the molder is shown in the act of venting with a vent wire.   Two rows of vent holes are shown.   These are used in cases where extra-good venting is required, as when the sand is of a fine close grain and has to be rammed hard,

FIG. 19.

or when it is wet.   When the sand is fairly open in texture, or when hard ramming is not necessary, and the sand is not too wet, one row of vents will be sufficient.   In some cases the venting is done with a single wire, which is forced down into the sand and drawn out at once as shown on the left of Fig. 19.   In other cases several rods are forced down and none withdrawn until all are in place, when they are drawn one by one. The advantage of the latter method is that the vents already made are not exposed to the danger of being crowded full of sand as the vent wire is thrust down near them. After the venting has been done, the vent channel at j is filled with cinders, the pattern is again banked with facing sand, and common sand then filled in for backing. Cinders are a necessity on some castings, but not always, and judgment is required in their use, as they increase the work of cleaning the sand for the next molding. This course is then rammed up in the usual manner, and so on until two more courses are completed, when another vent channel, as shown at j in the course marked 6, Fig. 17, is made and vented as above described. These courses of vents may be repeated any number of times until the top of the pattern is reached. Some molders lead the gases from these vent channels j into straight vent channels k at the joint of the mold. This is a bad plan, for if there is any breaking of joints by drawing the pattern or from a straining of the cope, allowing large fins or a run-out, the metal will run into the vent channels at k and fill them with iron; this may cause the loss of a casting by scabbing or blowing. In most cases of deep-sided molds, it is best to lead the gases from courses of cinder vents up to the joint by separate outlets, made by ramming up gate sticks in a mold, as seen at l, Figs. 17 and 18. These openings should be placed as far as possible from the face of the pattern and provision made for conducting the gases to them, as shown by the channel U, Fig. 17. To insure the best possible venting, the molder vents each course as it is rammed with a $\frac{1}{8}$-inch wire. These fine vents will give relief to the gases working through the sand to the large, or $\frac{1}{4}$-inch, vents shown leading to the channels j. Should these $\frac{1}{8}$-inch vents meet the pattern, the iron will not run into them and do damage as it would if the vents were $\frac{1}{4}$ inch.

30. Difficulties in Venting.—Where the sand can be worked fairly dry, or where it is of a very open texture so as to permit the gases to penetrate for a considerable distance, the method of venting illustrated on the right of Fig. 17 can be used with safety. This plan consists merely in using a $\frac{1}{4}$-inch vent wire, keeping it about 2 inches away from the face of the pattern, with the vents about 2 inches apart. In using the vent wire, it should be passed through the last two courses laid, and half way through the one under them. This is repeated at every other ramming until the joint on top of the mold is reached. In some cases a $\frac{1}{8}$-inch vent wire is used after ramming every course, and the $\frac{1}{4}$-inch vent wire after every other course, as just described. In case there is a cinder bed C, Fig. 17, under the mold, the vents may be made to connect with that bed, and may also be carried to the top joint.

31. Complicated or difficult cores are best made in dry sand, but they are sometimes made in green sand. Where difficulty is experienced in venting crooked bodies or long horizontal green-sand cores, the venting may be done with a special device. Instead of using the ordinary vent wire or rod, which is difficult to remove if long, and which cannot be removed if at all crooked, a vent rod is made by pushing a braided sash cord through a piece of rubber tubing and inserting the tube instead of the rod when making the core or mold. The act of forcing the braided cord through the tube pushes the cord together, making it somewhat thicker, which in turn expands the tubing and makes it practically solid. When this vent rod is to be removed it is done by first pulling out the sash cord, which allows the tube to collapse, so that it may be easily removed. Vents made in this manner may be carried long distances, and even around considerable bends that could not be dealt with by ordinary methods. The labor of pushing the sash cord through the tubing may be materially reduced by pouring a handful of fine graphite through the tubing before inserting the cord, and also rubbing the cord with the same substance.

32. Venting Shallower Molds.—Where molds are not over 20 inches deep, vents may be driven straight from the joint after the cope has been lifted, as shown by the vent wire in Fig. 18. This avoids any venting while ramming the courses. If there is a cinder bed under the pattern, the vents should be carried downwards to the cinders, and the top outlets k, Fig. 17, can be stopped up or dispensed with. This is a good scheme where there is any liability of the pattern breaking the joint and leaving a fin while being drawn, or any danger of the cope straining so as to let iron run into the vents when they are left open.

33. Using a Slanting Vent Wire. — In venting directly from the joint, care must be taken not to let large vent wires be driven in a slanting direction, as they may strike the pattern, as shown at vent v, Fig. 20, leaving a sharp point of sand at a; this may cause a slight blow at that point and burst the vent hole inwards, and thus cause a flaw that may spoil the casting. Such slanting vents are apt to fill with iron and cause scabs, or, worse still, cause a mold to start blowing. A mold would be far better without a vent than to have a lot of vents like those shown in Fig. 19 filled with iron, as in such a case they only help to create gas.

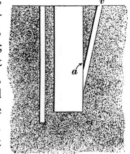

FIG. 20.

34. Cinders for Vent Channels. — In obtaining fine cinders for filling such vent channels as those shown at j, Figs. 17 and 19, the following plan is generally adopted: Cinders from an ash or coke pile are first placed in a $\frac{1}{2}$-inch riddle that is shaken to get the fine cinders and dust separated from the coarse material. The stuff that stays in the riddle is thrown away, and what has passed through is again riddled through a $\frac{1}{4}$-inch riddle. What passes through this second riddle is thrown away, and what stays in it are the cinders that are used for filling the vent channels at j. In obtaining such cinders, the two riddles can, if desired, be attached together and used at the same time, each being emptied when full.   The cinders thus obtained will be of a uniform and clean character.

EXAMPLE ILLUSTRATING DEEP MOLDING.

35. The Pattern. — The pattern shown in Fig. 21 illustrates a number of points that should be observed in

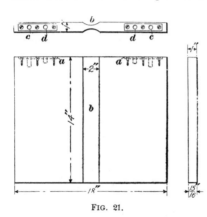

FIG. 21.

deep molding, and would be an excellent example for a beginner to practice on. It gives sufficient depth for three courses of ramming, and will afford good practice in learning to ram evenly. This pattern can be molded in a flask and then rolled over, as in Fig. 13, or it can be bedded in, as in Fig. 17. The pattern is made with a good taper to facilitate its being drawn from the sand.   This taper, or reduction in size of the pattern as it recedes from the face, is called **draft.**   For convenience in handling the pattern while drawing it from the sand, iron draw-plates are screwed to the pattern, as shown at a, a, in which are the rapping holes d, d, and the screw holes c, c for the draw-screw. The groove b provides a little irregularity in the pattern that will the more thoroughly test one's ability to ram evenly.   Castings of the form of Fig. 21 are sometimes molded with one of the long dimensions vertical, in order to get smooth castings.

36. Making the Drag.—The principles involved in ramming and venting the sides of deep patterns have been described and illustrated in Arts. **25** to **34,** and the detail

of ramming up the drag will therefore be omitted.   Fig. 22
shows the drag rammed up, the bottom board bedded on

FIG. 22.

solidly, and all clamped together ready to roll over.   Fig. 23
shows the flask rolled over, the molding board removed, and

FIG. 23.

the molder making the joint by *sleeking* it over firmly with
a trowel.   After the joint is thoroughly sleeked so as to

give the sand a fine finish, a brush is used to free the surface of all dust.

37.   Parting Sand.—The joints having been thus finished, parting sand is shaken from the hand, as seen in Fig. 24, in such a manner as to distribute it evenly over the

FIG. 24.

joint.   The use of parting sand is to cover the joint with some material that is not adhesive, so that when it is spread between the bodies of damp sand, it will allow them to separate without pieces of one body adhering to the other..   Material for making parting sand is generally obtained in foundries from the fine dry sand that sticks to the surface of castings before they are cleaned.   All the clay contained in this sand has been burned hard, and for this reason when placed between two damp bodies of green sand it will prevent their sticking together.   Another kind of parting sand is obtained by using fine grades of lake, river, or seashore sands that have been dried on a plate over a hot fire.   Before using this shore sand or the dry burned sand from castings, it is passed through a sieve as fine as can be used—in some cases a flour sieve. After the parting sand is placed on the drag in such cases as Fig. 21, where there are cross-bars in the cope, a small amount of facing sand is sifted over the parting sand and cleaned off around the joints of the flask.

38. Ramming the Cope.—After the joint is made and covered with facing sand, the cope is put on and the pins or patterns for forming the gates are set ready for ramming, as shown at *e* and *f*, Fig. 25. While ramming

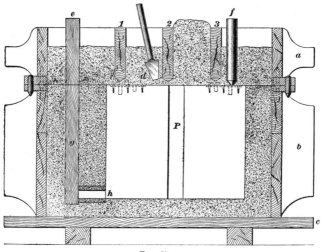

FIG. 25.

the drag, the pattern *g* and the hollow core *h* for the thin rectangular gate were put in place, as shown in Fig. 25. This form of gate is used to introduce the metal near the bottom of the mold. Before the cope is put on, its bars *1*, *2*, and *3* are thoroughly wetted with water, to make the sand stick to them. If the sand is not of such a loamy nature as to hang well, a clay or loam wash or a thin flour paste is often used instead of water on the face and sides of the bars. After the cope is in place, a riddle is used in filling the cope to a depth of about 2 inches. This. operation is followed by tucking the sand under the bars with the fingers of each hand, until the sand is as solid as the fingers can ram it. Unriddled sand is then filled into the cope to the level of the top of the bars. The peen end of the rammer is now used to ram the sand firmly between the various bars, which will pack the sand to about the level seen between bars *1* and *2*. It is very important to peen the first course thoroughly. If this is not done, there may be soft spots in the face of the mold under the bars; or the sand at the bottom of the bars may be so soft as to drop out when lifting off or closing the cope. After the first course of sand is thus peened between all the bars, more common sand is filled in and heaped above the level of the bars, as seen between *2* and *3*. This done, the peen is again used to ram the second course of sand between all the bars. Less care and time can be spent in peening this second course, because the butt of the rammer is to follow the second peening. In peening the first course, the peen should be directed in such a way as to pack the sand under the bars, as seen at *d*. When both courses have been peened, the butt of the rammer is used to ram down the mounds of sand between the bars, as between *2* and *3*. The butt is struck down solidly to pack the sand between the bars. With the first course it is important that it be firmly peened to keep it from dropping out, and this is also true with regard to the butting, for if that is not done firmly, the cope is liable to drop out.

In ramming the cope, care must be taken not to strike the bars, as that might loosen the sand and cause it to drop out when lifting off the cope or closing the mold. After the cope has been rammed, any excess of sand is removed by striking off the top of the mold with a piece of board or with the handle of the floor rammer. The mold is then vented with a $\frac{1}{8}$-inch wire, the vents being about 2 inches apart all over the area covered by the pattern. This done, the sprue pin *e* in the pouring gate and riser pin *f* are withdrawn. Then the tops of both gate and riser are reamed out funnel-shaped, and all loose sand firmly tamped down with the fingers and then dampened lightly with the swab, so that in pouring the metal, it cannot wash any dry dust or sand into the mold along with it. The mold is now ready to have the cope lifted off, which is done, and the cope placed on any suitable support, as the box shown in Fig. 26.

FIG. 26.

39. Venting the Drag.—After the cope is lifted off, the parting sand is carefully brushed off the joint and the joint swabbed, care being taken not to get too much water on it around the edges of the pattern. This being done, a $\frac{1}{4}$-inch vent wire is used to vent the sides opposite the places marked h, h, h, Fig. 26, and a groove is run from each vent to the outside of the mold. It will be noticed that this method of leading away the gases from the joint differs from that illustrated in Fig. 18.

In the case of molds having joints that are liable to be broken in drawing the pattern, the method of venting just described should be used, because if the iron should get in the joint at any one point of a vent channel, as at k, Fig. 17, it might readily fill all the vent holes.

40. Drawing the Pattern and Finishing the Mold.—After the mold is vented, the pattern is loosened by **rapping.** This is accomplished by placing a pointed iron bar in the rapping holes shown at *d, d,* Fig. 21, and rapping it with an iron bar or a hammer. Where this is done, lifting screws are placed in the holes *c, c* of the draw-plates and two men gradually start the pattern, each of them holding a lifting screw in one hand and a hammer in the other, with which they tap the pattern lightly on the draw-plates *a, a,* Fig. 21. Care must then be taken to bring the pattern up steadily until it is out of the mold. After the pattern has been drawn out, the gate *g,* Fig. 25, is drawn. The trowel is next used to press down any portion of the joint that may have been started in drawing the pattern, to smooth up the cope part of the mold. The top of the pouring gate and riser are then reamed out funnel-shaped. Should any dirt have fallen into the mold, a lifter is passed into the mold and the dirt removed.

41. Closing and Casting the Mold.—The mold is now closed and clamped with two clamps, as shown at *i* and *j,*

Fig. 27. It is now ready for casting, which is done by two men holding the ladle and pouring the mold, while a third one is skimming. Molten iron always contains more or less scum or dirt, which, if allowed to pass into the casting, will make flaws that will impair its strength and finish, and may even

FIG. 27.

cause its loss. For this reason when molten iron is being

poured from a ladle, a skimmer should always be used to hold back the dirt.

42. Shaking Out the Casting.—About 30 minutes after the casting has been poured, it can be **shaken out,** which means that the cope may be taken off the drag and the drag itself removed from the mold, care being taken not to disturb the sand about the casting. After some hours the casting may be removed from the sand, when, with the gate and riser attached, it will appear as in Fig. 28. The thin rectangular gate is shown at *g*, the vertical por-

FIG. 28.

tion of the gate at *e*, and the riser at *f*. The pouring gate can be freed from the casting with a fair blow of a hammer. When the gate is removed and the casting cleaned, the job is finished. By measuring the casting and comparing it with the pattern, the molder can tell how well his ramming prevented the distortion of the casting due to the pressure of the metal in the mold.

43. Allowance in Pattern for Distortion.—While it is true that hard ramming tends to decrease the amount of distortion that occurs in the lower part of deep molds, it is nevertheless impracticable to ram very deep molds hard enough to prevent more or less distortion. Few experienced molders ram alike, and some produce castings more distorted than others. No definite rule can be given for the allowance to be made for distortion in deep castings due to the straining of the mold by the increase of pressure in the lower part of the mold. In the case of experienced molders who are recognized as hard rammers, an allowance of $\frac{1}{32}$ inch per foot would be sufficient, while equally experienced men who do not ram their molds as hard might require an allowance of $\frac{1}{16}$ inch per foot. In the case of a pattern 5 feet deep, if an allowance of $\frac{1}{16}$ inch per foot were made, it would decrease the dimension of the pattern at the bottom of the mold $\frac{5}{16}$ inch. From this it will be seen that the allowance for distortion tends to give draft to the pattern. Where the conditions are all known, it is possible to produce practically parallel castings from a tapered pattern, the distortion due to the straining of the lower portion of the mold just neutralizing the amount of draft necessary on the pattern.

SHALLOW MOLDS.

44. The Horn Gate.—In Fig. 29 a shallow mold is shown to illustrate the use of a horn gate, which is shown at *b*. The pattern *a* is of a small toothed gear-wheel, and

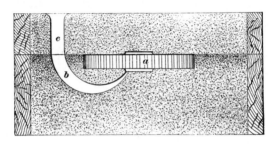

FIG. 29.

the horn gate *b* connects the lower boss with the gate *c* in the cope. A horn gate is only used where the circumference of a small casting is of such a delicate nature that it does not afford a place to gate the casting on the side, such as the gear-wheel shown in the illustration. If this should be gated at the center on top, there would probably be danger of the iron splashing over and knocking down some of the teeth. For this reason the horn gate *b* is used and the casting is made without great difficulty. A gate of this kind can be readily removed from the casting on account of the small connection with the casting.

45. Ramming a Flat Surface.—The molding and casting of a square, flat plate, about 10 inches square and

1 inch thick, will be considered next.   In ramming flat or
plate surfaces like the pattern shown at P, Fig. 30, the peen

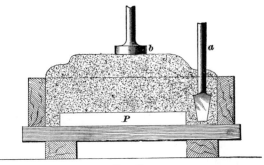

FIG. 30.

a should never be used over the face of the pattern during
the ramming of the first course; if this is done, it causes
uneven ramming and may
make hard spots in the
mold, and so cause scabs
or indentations that may
spoil the casting.   In the
case of flat or plate work,
the sand is rammed up in
courses from 4 to 5 inches
thick.

Besides ramming with
the floor rammer a and
b, as shown in Fig. 30,
it is very common for
the molder to ram with
his feet, as shown in
Fig. 31.   After peening
around the mold next to
the flask, he steps on top of
the sand heaped up on the
middle of the mold and

FIG. 31.

moves across in a series of short jumps, keeping both feet
together.   He then steps to one side and treads across again,

jumping in the same manner. This is done until the molder has covered the whole area of the top of the mold, when he steps on the ground, rams the surface with the butt of the rammer, and strikes off the top of the mold.

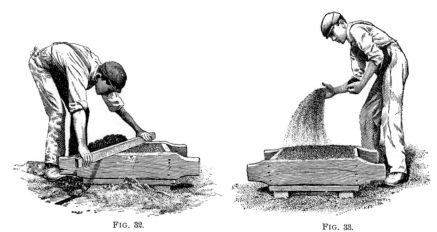

FIG. 32.   FIG. 33.

The operation of striking off the drag is shown in Fig. 32. The sand having been struck off as shown, the hand is used

FIG. 34.   FIG. 35.

to sprinkle molding sand to the depth of about $\frac{1}{4}$ inch all over the surface of the struck-off sand, as shown in Fig. 33. This is done to allow the bottom board to be rubbed down to a solid bearing when in place, as shown in Fig. 34.

46. Venting the Drag.—After the bottom board has been bedded on solidly, it is taken off again and a rod, rammer, or strike is used to form creases or indentations in the sand, as shown at a, Fig. 35. These creases are placed about $1\frac{1}{2}$ inches apart and permit the gases from the $\frac{1}{8}$-inch under vents to escape along the face of the bottom board when the mold is being poured. The vent passages having been made, as at a, Fig. 35, a $\frac{1}{8}$-inch vent wire is then used, in the manner shown at b, to vent the drag all over the surface covering the pattern. The vent wire should not be driven down until it strikes the pattern; if it comes within $\frac{1}{2}$ inch of the face of the pattern, it is sufficient.

FIG. 36.

The entire area over the pattern having been vented, the bottom board is replaced and clamped, so that the nowel may be rolled over, as shown in Fig. 36.

FIG. 37.

47. Venting the Cope. The drag being rolled over, the joint is prepared as previously described. The cope is then put on and the sprue pins for forming the pouring gate and riser set in place, as shown at e and f, Fig. 37. After this, the cope is filled with heap sand, rammed as described in Art. **38,** and vented over the part above the pattern with a $\frac{1}{8}$-inch vent wire. It is then lifted off and finished.

48. Swabs and Swabbing.—The cope having been finished, the joint is brushed off and dampened with the **swab** shown in Fig. 38 (*a*). The swab is dipped in water

FIG. 38.

and then lightly squeezed out, after which it is passed over the joint, as shown in Fig. 39, care being taken not to get the sand too wet along the joint between the pattern and sand. Where delicate swabbing is required, as in bench molding, or the wetting of small bodies of sand, a sponge with a wire or long thin nail passed through it, Fig. 38 (*b*), is often used, as shown in Fig. 40. The sponge holds the water until squeezed out with the hand, and the wire or nail directs the water to the spot to be dampened. In making the swab shown at (*a*), a piece of fine-grade, long-fiber hemp rope is

FIG. 39.

tied at one end with a fine wire or string to form a handle, and the other end *teased out* by pulling it through a comb made by driving some nails through a board and allowing the points to extend beyond the board. While in use, the board forming the comb may be fastened to a bench or table. In making such swabs, some molders use a fine string that will not unravel in water or when dampened.

FIG. 40.

This string is cut into lengths of about 1 foot, and a bunch of them tied at one end to form a handle, as described. Good swabs can, however, be obtained very cheaply from any foundry supply house.

49. Rapping and Draw-Plates.—After the pattern has been swabbed, it is rapped by pounding on the side of a bar that has been inserted in the hole in the rapping plate, as shown in Fig. 41. It is better, however, to have two holes in the rapping plate, one for the rapping bar, and the other for the draw-screw. Some persons make the draw-screw serve for both purposes; this may answer where the patterns are light, but for large patterns there should be a rapping hole as well as a draw-screw hole. If the screw hole is made to serve for the rapping hole also, the thread in the screw plate is soon destroyed, even if the rapping is done on the draw-screw. Too much care cannot be taken to have good arrangements for rapping and drawing patterns, for this not only prevents the wear of the pattern,

FIG. 41.

but also saves labor in molding. After the pattern has been loosened by rapping, it is drawn by introducing the drawhook or draw-screw, as shown in Fig. 42. While drawing the pattern, it should be rapped gently, as shown in the figure. The rapping plate on the pattern shown in Fig. 42 is made large enough to contain two holes, one for the draw-screw and one for the rapping bar.

FIG. 42.

50. Cutting the Gates and Riser. The pattern having been rapped and drawn, the mold is finished with the trowel and the double ender, the latter tool being shown at (*b*) in Fig. 43. The

" Yankee," shown in Fig. 43 (*c*), is used for removing any loose sand from the mold or for finishing corners. The gates for connecting the sprue and riser with the mold are cut by means of a gate cutter. The form of gate cutter illustrated at Fig. 43 (*a*) consists of a thin blade of brass or steel bent to the required form and provided with a handle. Another form of gate cutter is shown in use in Fig. 44. This form of cutter consists of a thin sheet of tin or

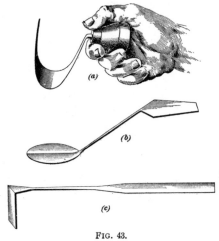

FIG. 43.

brass bent to the required form and is used as shown in the illustration. After the gates *f* are cut to the bottom of the sprue or riser *e*, Fig. 44, they should be smoothed firmly

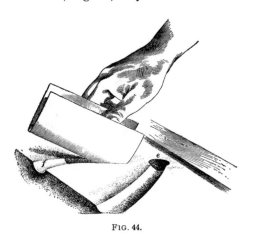

FIG. 44.

with the fingers, so as to press down all the loose sand. If this is not done, the iron, when running through them, will carry dirt into the casting. Before smoothing down the pouring gates, it is well, with some molding sands, to wet the top edge of the gates by carefully using a swab or sponge, as shown in Fig. 40. In using water in this way around gates, extra care must be exercised not to make any portion of the sand too wet; for this might cause the iron to blow as it passed into the mold and result in spoiling the casting.

FIG. 45.

51. Closing and Pouring the Mold. The mold having been finished, it is closed, and the flask

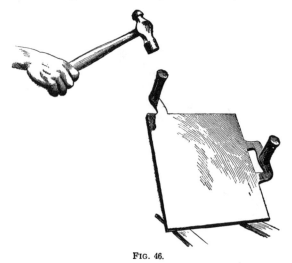

FIG. 46.

clamped ready to be poured, as shown in Fig. 45. In

20 minutes after the casting is poured, the cope and drag may be removed. After a few hours, or on the following day, the casting may be taken from the sand, when it will appear as in Fig. 46, which shows how the pouring gate and riser are connected with the casting. After the casting is cool, the gate and riser are knocked off by means of a hammer, as shown in the upper left-hand portion of the illustration, and the casting is then cleaned up, thus finishing the work so far as the foundry is concerned.

GREEN-SAND MOLDING.

(PART 2.)

MOLDING BY BEDDING IN.

LEVELING.

1. Using Straightedges.—Having described the casting of flat surfaces in green sand by rolling over, we shall now deal with the making of a casting by **bedding in.** A bed for a flat plate may be made at the top of the molding floor or at any desired depth below it. In any case it will be necessary to have some guide to form the bed; this guide generally consists of straightedges, arranged as shown at *a*, *b*, and *c*, Fig. 1.

In leveling up the straightedges, the first one *a* has a mound of sand placed under each end, so as to keep its under edges free from the floor, and is made as nearly level as the eye can judge. A spirit level is then placed on it and the high end driven down with a wooden mallet or a hammer, the pounding being done on a block of wood (to prevent the hammer marring the face of the straightedge), as in Fig. 2. The straightedge *b* is then placed on two mounds of sand, as was done with *a*, setting it by eye as nearly true with *a* as possible. The straightedge *c* is then placed on the ends of *a* and *b*, Fig. 1, and the spirit level used to make the ends of *b* level with *a*. This can be done by first bringing one end of *b* to the level of *a* by using the straightedge *c* as shown, and then carrying *c* to the other end and repeating the operation. Still another plan to get *b* level with *a* is to use *c* as

§ 41

shown, and then remove the spirit level from *c* to *b* and raise or lower the other end of *b* as may be found necessary.

It must be remembered that both edges of the straight-edge *c* must be parallel; this is not necessary, however, in

FIG. 1.

the case of *a* and *b*, the upper edges only of these two being required to be true. These straightedges should have a hole in one end so that they can be hung up when not in use.

FIG. 2.

They should not be left lying around on the floor, where they are liable to have their edges injured or to be bent or warped from exposure to dampness and uneven supports.

MAKING THE BED.

2. General Remarks.— The straightedges having been leveled as described, the work is proceeded with according as the thickness of the casting to be made requires the bed to be hard or soft.   In pouring castings in **open sand,** that is, pouring flat plates without a cope covering, a soft bed is generally used.   The bed, when down or under vents are not used, must be soft in order to permit the gases to escape freely from the sand, for there is no head pressure of metal on the bed to drive out the gases as there is when a plate casting is poured through a cope.   In the latter case, the bed is generally made hard in order to withstand the head pressure of the metal.   Much more labor is required in making a hard bed than a soft one.

3. Making a Soft Bed.—In making open-sand plate castings, the bed must be softer for a casting from $\frac{1}{2}$ inch to 1 inch thick than for one whose thickness is from $1\frac{1}{2}$ to 3 inches.   The reason of this is that the thicker the casting, the greater is the pressure exerted by the metal on the bed, tending to drive the gases downwards into the lower part of it or to cause them to escape outwards at its sides. The same degree of softness necessary for the thin plate may sometimes be used for a thick one ; but the harder the bed can safely be made, the better, as this prevents the pattern, or any other light body, from making impressions on it and causing an uneven face on the casting.   It is well to have a bed as hard as practicable, still it is better to have the bed soft than too hard.   By using care in smoothing soft beds with the trowel and in laying on the patterns, a casting can be produced with almost perfectly true surfaces.

In starting to make a bed after the straightedges have been leveled up, sand is tucked solidly under a and b, Fig. 1, so as to prevent the pounding action of the straightedge c from disturbing their level.   After this is done, sand is shoveled on both the inside and outside of a and b until it is level with their tops.   The sand on the outside is then packed down with a butt rammer or with the feet, to prevent the pressure of sand on the inside of *a* and *b* from moving them outwards. Next, the straightedge *c* is used to strike off the sand level with the top of *a* and *b*. Sand that has been passed through a ¼-inch riddle is now spread over the face of the bed to a depth of about ¾ inch, and then a flat piece of wood or iron about 8 inches long is placed on each of the straightedges. These pieces *d* and *e*, Fig. 3, should be

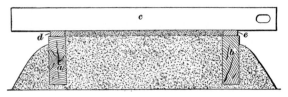

FIG. 3.

from ¼ inch to ½ inch thick, according to the desired hardness of the bed; the thicker the piece, the harder will be the bed produced. On top of *d* and *e* is laid the straightedge *c*; a man is then stationed at each end, holding the pieces *d* and *e* and also one end of *c*. These men then pull the straightedge *c* and the pieces *d* and *e* along, sweeping to the end of the bed all sand that lies above the level of *d* and *e*. This leaves a body of sand, the thickness of the pieces *d* and *e*, projecting above the level of the straightedges *a* and *b*, which sand is next pounded down to the level of *a* and *b* by means of the edge of *c*. To do this, a man holds one end of *c* down on one of the straightedges *a* or *b*, while another man raises the other end from 4 to 6 inches and brings it down upon the raised sand so as to drive it down to the level of the top of *a* or *b* on his side, as shown in Fig. 4. When one man has gone a distance of about 12 inches, he returns about half way to his starting place and holds his end of *c* down on the under straightedge, *a* or *b* as the case may be, while the opposite man raises the other end of *c* and pounds down the sand on his side. This operation is continued by the two men alternately, until the whole surface of the raised sand is pounded down to a level. When the bed has been thus gone over, its surface will present a series of imprints of the straightedge c.   To remove these marks, the straightedge c is dragged in a seesaw manner across the

FIG. 4.

face of the bed, as shown in Fig. 5, each man being careful to have every stroke a forward one.   If this were not done, marks caused by the backward motion of the straightedge

FIG. 5.

would be left on the surface, while the portion that was done properly would be smooth.   The portion of the bed shown in Fig. 5 that is nearest the straightedge has been done properly, while the portion near the front of the illustration has been struck off with alternate forward and backward movements of the straightedge so as to leave ridges of sand on the bed.

To accomplish this forward movement requires a little practice, as it is rather difficult to have every movement a forward one while seesawing the face of the bed. If the face of the bed is struck off by a *straight pull*, not only will the surface be left rough, but the operation is apt to loosen the face of the pounded sand from the underlying soft body, so that when the metal runs in over the sand, it is liable to loosen or lift the sand in such a way as to cause a rough and scabby casting. After the bed has been struck off as described, a fine sieve is used to shake a very thin coating of sand evenly over the face of the bed. A trowel is then used to smooth down the sieved sand, after which the bed is ready and the molder can proceed to build up the sides of the mold with sand, and so give any shape required in the

FIG. 6.

casting. A clear idea of this may be gained from Fig. 6, which shows a man packing sand against the side of the pattern *P* in such a manner as to form the sides of the mold.

The sides having been built up as described, a pouring basin *g* is made, and the mold is ready to have the metal poured into it. For pounding down with the straightedge some molders use a sharp grade of bank or river sand, alone or mixed with common heap sand.   This is only necessary where the molding sand proper is of a close, fine grain, as such fine sand is liable to cause scabs on account of its not being sufficiently open to permit the gases created at the face of the bed to escape downwards freely.

Where sharp sand is used to form the face of the bed, it may be omitted in front of the pouring basin, and this place filled in with a stronger and closer-grained sand, as shown at the light spot marked h, Fig. 6.   For this purpose some molders use ordinary molding sand, or a regular facing sand, which is made by mixing sea coal with molding sand, in front of the pouring basin, in place of the ordinary sand just mentioned.   The reason for using a stronger grade of sand in front of the pouring basin is that the metal, when flowing from the basin on to the face of the bed, is liable to break the latter where the sand is weak and cause it to be washed in along with the metal, the result being a rough or scabbed face (on the casting) in front of the pouring basin.

The subject of venting open-sand beds will be treated further on.

4.   Thickness of Open-Sand Castings.—The thickness of open-sand castings is determined by the thickness of the piece P used in building up the sides of the mold, Fig. 6, by the fluidity of the metal when pouring the casting, and by the judgment of the molder in deciding when sufficient metal has flowed into the mold.   An iron that is *dull* or not very fluid will readily pile up in an open-sand mold, so as to be from $\frac{1}{8}$ inch to $\frac{1}{4}$ inch thicker than the height of the sides of the mold, for which reason it is difficult to obtain open-sand castings within $\frac{1}{8}$ inch of any specified thickness.   To attain even this accuracy, the metal used should be very fluid.

5.   Venting Soft Beds.—When soft beds are made as here described, and the castings (where more than one is placed on the same bed) are from 8 to 12 inches apart, they will seldom require any venting, except when it is necessary to use fine grades of sand.   In such cases vents may be made as described further on.   Where floor space is limited so that open-sand castings must be placed close together, say from 2 to 4 inches apart, it may suffice, where open grades of sand

FIG. 7.

are used, to drive a row of vertical ⅜-inch vents down the center of the partitions of sand dividing the molds, as shown at *v*, Fig. 7.   There should never be any difficulty in carrying off gases from soft beds when it is desired so to do.

6. Making a Hard Bed.—There are few things in founding in which molders differ so much as in their methods of making hard beds.   It may be said that there are four degrees of hardness in making beds: The first is the soft bed for open-sand castings of plates ranging from ½ inch to 3 inches thick; the second bed is of the character required for making the prickered plates used in loam molding; the third is for hard beds used for open-sand castings that must be well vented; and the fourth is for beds that are used for light and heavy castings in cases where the bed forms the bottom

part of a mold that is covered with a cope. The most difficult beds to make are those used for making loam plates
requiring long prickers on them, as shown in Figs. 7 and 8,
and also in the matter treating on loam molding.

7. Making Beds for Prickered Plates.—In making beds for plates having prickers, or prongs, on them, the
depth of the tempered sand must be much greater than for
plain plates. As an example, the plain plate to be cast in
the bed shown in Fig. 6 can be successfully made with a
depth of tempered sand ranging from 4 to 5 inches without
any venting; but if it should be necessary to have prickers
from 4 to 6 inches long on such a plate, the tempered sand

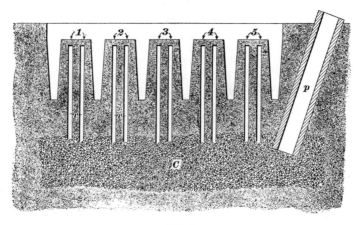

FIG. 8.

should be fully from 7 to 10 inches deep. This is due partly
to the necessity for packing the sand firmly with the palm
of the hand or the back of a shovel as every course of sand
of from 3 to 4 inches deep is filled in. If sand were shoveled
in loosely, as in the case of the soft bed just described, the
prickers or prongs, if over 6 inches deep, would be liable to
be strained at their bottom ends in such a manner as to burst
into each other, or to swell at the end so badly as to raise
the part of the mold above the lower end of the prickers,

thus spoiling the casting. Some loam plates require prickers or prongs from 12 to 30 inches deep. It is very difficult to make such deep-prickered plates in open sand without their blowing more or less when being cast. In some cases it is best to dig deep enough to permit a cinder bed being placed under such plates, as shown at C, Fig. 8, and then filling in the tempered sand in layers ranging from 3 to 4 inches deep, and venting each alternate layer of sand with $\frac{1}{8}$ or $\frac{3}{16}$-inch vent wire down to the cinder bed, which is itself vented through vent pipe p. Another plan is to have six or ten pricker patterns instead of the one or two used in the

FIG. 9.

case of shallow prickers, and to press them all down one after the other, and then, before they are withdrawn from the sand, take a $\frac{1}{8}$-inch or $\frac{3}{16}$-inch vent wire, as seen at i, Figs. 7 and 9, and carefully vent down from the face of the mold to the cinder bed between all the pricker patterns, making the vents about 2 inches apart, as at v, v, Fig. 8. These vents, coming to the surface of the mold, must be stopped up with the end of the finger and then all the finger holes filled with sand and the surface sleeked over to make the face of the mold smooth, as at 1, 2, 3, 4, 5, Fig. 8.

In some cases, instead of carrying the gases from these vertical vents through the cinder bed, the venting can be done by using a long $\frac{1}{4}$-inch or $\frac{3}{8}$-inch vent rod driven in at the bottom of the tempered sand to bring the gases to the side of the bed, as seen by the vent wire k, Figs. 7 and 9. In packing the sand by layers or courses in these beds, care must be taken not to make them so hard that the molder cannot push the pricker pattern down by hand to a depth of from 6 to 8 inches. Beyond this depth it may often be necessary to use a hammer to drive the prickers, as in Fig. 7. Another point to be noted in making deep-prickered plates is never to use any sand but that which has been well tempered and passed through a $\frac{1}{4}$-inch or $\frac{1}{2}$-inch riddle.

Care must be taken to work the sand as dry as it can be used and give good results; it must not be so dry, however, that there will be any risk of the sand running back into the holes left by the pricker pattern after it has been withdrawn. The pattern used for making prickers or prongs on plate castings should always have a handle from 6 to 8 inches long and of convenient form for the hand. A good form of pricker pattern is illustrated in Fig. 10, where it is shown as generally held when being pressed into the sand.

Deep-prickered plates usually give trouble at the first attempts, but with some experience in molding them, and by following the directions here given, they should be made successfully.

8. Hard Beds for Open-Sand Castings.—Castings with lugs or long projections on them may be cast in open-sand molds if it is not of much importance whether the cope, or upper side, of the casting is a little rough. In making such

FIG. 10.

castings, the straightedges are leveled as already explained.

Then, tempered sand is filled in and rammed down evenly and firmly to nearly the top of the straightedge, using the butt of the rammer or the feet. When this is done, a straightedge c, Fig. 9, with its lower edge cut down $\frac{1}{2}$ inch at each end, is used to strike off the bed so that its surface will be $\frac{1}{2}$ inch below the top of the straightedges a and b. Then the surface of the bed is vented all over with a $\frac{1}{8}$-inch or $\frac{3}{16}$-inch vent wire, the vents being made about 1 inch apart. These vents may be carried down to a cinder bed or led from under the bed with $\frac{1}{4}$-inch or $\frac{3}{8}$-inch vents leading from under the straightedge, as shown by the vent wire projecting at k. After the surface of the bed has been well vented, the palm of the hand is passed over it to close up the top of the vent holes. Riddled sand is then shoveled on to about $\frac{1}{2}$ inch higher than the top of the straightedges a and b. Pieces d and e, about $\frac{1}{4}$ inch thick, are now used in conjunction with the straightedge c, to strike off the bed as shown in Fig. 3. When this is done the surface of the bed is pounded down with the straightedge c and the face of the bed struck off in a seesaw manner and finished as already described.

MOLDING PLATES WITH PROJECTIONS IN OPEN SAND.

9. General Directions.—The hard bed, Fig. 9, having been completed, it may be used for casting plates having

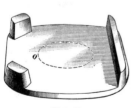

FIG. 11.

projections, flanges, lugs, etc., of which the pattern shown in Fig. 11 is a type. In bedding such a pattern on a level bed, it is first set on the bed and pressed down so as to leave an imprint of its projections; it is then removed and the shovel used to cut away the greater part of the sand where the lugs come, after which a trowel is used to cut away the sand to within about $\frac{1}{4}$ inch of the inside of the imprint. This leaves the bed as shown in Fig. 12. The pattern is next set back and hammered down solidly on the bed, the molder striking on a block, as shown in Fig. 13; after this, sand is rammed up against the outside of the lugs and the edges of the pattern and the mold vented, as shown at *k* in Fig. 14. Next the pattern is drawn and the mold finished ready for the metal, as shown. It is of special importance that a block of

FIG. 12.

wood or a heavy plank be used to pound on when bedding in patterns; no sledge or heavy hammer should be used on the bare face of a pattern.

FIG. 13.

Another method of making a casting by bedding in, from a pattern like that shown in Fig. 11, is to have a hole cut in the pattern about the size shown by the dotted line at o, and then to set its projecting lugs on mounds of sand and tuck up the pattern on the inside through the hole o, as shown in

FIG. 14.

Fig. 15. This answers well where the patterns are not of large area, but for large plate surfaces with patterns having flanges or lugs, the plan shown in Figs. 12 and 13 is a good

FIG. 15.

one. One objection to this plan, however, is that the projections, flanges, or lugs are faced with common heap sand. In order to obtain a smooth surface throughout, such projections as shown in Fig. 11 should be faced on the inside with sand mixed with finely ground coal, commonly called *sea coal*.

10. Facing Inside of Projections.—To face the inside of projections, flanges, etc., when the pattern has been bedded in the manner shown in Figs. 12 and 13, the pattern must be drawn before the outsides of the projections are rammed up, and then a section of the inner face of the projection cut out, as shown at *a*, Fig. 16. A piece of board or

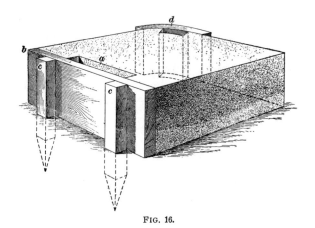

FIG. 16.

plank must next be laid against the projection, as shown at *b*, and then the cavity at *a* rammed up with facing sand that has been mixed with sea coal. A $\frac{1}{8}$-inch vent wire is used to vent the face and sides of the projection, keeping the vents about 1 inch back from the face of the board *b*, and driving them to the cinder bed or horizontal vents, as shown at *k*, Figs. 7, 9, and 14. When this is done, the tops of the vent holes are stopped up with the finger and then filled with sand, and the face of the mold finished with a trowel, after which the board *b* is removed. If the section *a* is not long enough to cover the entire inner face of the projection, another section joining the portion just faced is cut out and the board *b* again replaced and the operation repeated as described. This operation is repeated until the whole face of the projection has been lined with rammed facing sand from 1 inch to 1½ inches thick. The length of *a* will depend on the length of the surface to be covered with facing sand; this can extend from 4 inches to 4 feet, so long as the board *b* is kept in the required position. To hold such boards, or **formers,** as *b* in the right position while the cavity at *a* is being rammed, stakes *c* are driven into the floor. The space or cavity at *a* is rammed with ½-inch to ¾-inch rods, or with a small hand rammer having a peen 1½ inches wide and ½ inch thick. Care must be taken to face and ram the space at *a* evenly and firmly. In the case of projections having forms that are not straight, it will be necessary to use special pieces or patterns, as shown at *d*, the whole being rammed, vented, and finished the same as if the piece were straight. Boards used as formers *b* must be smooth on the side forming the face of the projection.

HARD BEDS FOR COPED MOLD.

11. Making the Bed.—Large surfaces covered with a cope permit of harder beds being used without causing the metal to blow and create scabs than can be used for open-sand work. In making these beds, it must be remembered that the sand should be as dry as can be easily worked, and should be vented as well as possible. The operations, so far as leveling the straightedges is concerned, are nearly the same as those described in Art. **1,** the only difference being that the straightedges *a* and *b* will be set down into the floor about two-thirds of their depth, instead of being set up above the floor, as seen in Figs. 1 and 2. After the straightedges are leveled, sand that has been well tempered and riddled is filled in between them and struck off level with their top edges; the sand is then rammed·down with the feet as far as possible, and then the whole surface of the bed is rammed with the butt of the rammer. This will bring the rammed sand down to about 2 inches below the top of the straightedges, when these are about 6 inches deep, which is about as deep as such a course should be. The sand having been butted solidly, common sand is again shoveled in and struck off to the level of the top of the straightedges. This is then butted down lightly all over the bed and a straightedge c, Fig. 9, whose ends are cut down $\frac{1}{2}$ to $\frac{3}{4}$ inch, is used to strike off the bed, which is then closely vented all over the surface with a $\frac{1}{8}$-inch or $\frac{3}{16}$-inch vent wire, and the top of the vents stopped up with the palm of the hand; next, sifted sand, which for some classes of work may be mixed with sea coal, is shoveled in and struck off with the use of $\frac{3}{8}$-inch strips on the straightedge a and b, Fig. 3, as described in Art. **3,** and then pounded down and finished, as already described.

12. Objectionable Method of Venting. — Some molders make a practice of venting direct from the face of the bed and then closing each hole with the finger to stop the surface of the vents, instead of having the tops of the vents about $\frac{3}{4}$ inch below the surface of the bed. After the upper ends of the vents are all stopped with the end of the finger, facing sand is packed solidly into each hole with the fingers, and the bed is then smoothed with a **surface block** (preferably of hard wood), measuring about 3 inches by 8 inches, with the corners a little rounded. This plan will succeed if carefully carried out, but should the top of a few holes be carelessly stopped up, there is danger of the iron bursting through and running down the vents to the cinder bed, if there is one. Then, again, gas might become confined in the cinder bed and, by exploding, create an upward pressure in the vent holes and lift the sand from the top of those vent holes that have not been filled firmly with sand. A little study of this method of venting will make its objectionable features evident.

13. Surface Venting. — Where beds are desired with very hard surfaces, or where the sand is of such a nature as to scab readily, it is a good plan to bring the vents as close to the surface as practicable. The following is a plan for surface venting that, if followed, will dispense with the finger poking of the vent holes and will insure hard beds being reliably vented. The plan consists in ramming up solidly and striking off the common sand within about $\frac{3}{4}$ inch of the top of the straightedges, Fig. 9, and then covering the surface of the bed with facing sand mixed with sea coal, so that it projects about $1\frac{1}{4}$ inches above the top of the straight-edges, using pieces $1\frac{1}{4}$ inches thick under the straightedge c,

FIG. 17.

as shown at d and e, Fig. 3. After this is done, a butt rammer is used over the entire surface, in the manner shown in Fig. 17. In butting the surface of such a bed, the operator must make sure that no part of the surface is missed, and the butt rammer must be applied lightly so as not to make the face of the bed too hard. To insure a smooth surface to the bed, no sand should be allowed to stick to the face of the butt when ramming; a brass butt is good for such work. The sand, after being butted, should project above the straightedges about $\frac{1}{2}$ inch. The bed having been butted all over, its surface is vented with a $\frac{1}{8}$-inch wire, placing the holes about 1 inch apart, and driving the vent wire down to the cinder bed, should there be one under it. If there is no such bed, then $\frac{1}{4}$-inch or $\frac{3}{8}$-inch vent rods, set horizontally, as seen at k, Figs. 7, 9, and 14, will have to be used. After the venting, the bed is struck off in a seesaw manner, as already described, and facing sand sifted all over its face as thinly as possible. This done, a hard-wood finishing block is worked all over the face of the bed to make it smooth and ready for finishing with the trowel. The application of the smoothing block not only gives a finish to the surface of the bed, but it also assists in firmly stopping up the top of the $\frac{1}{8}$-inch vent holes. This not only prevents any iron from bursting into them, but also prevents any confined gas from forcing its way through the holes into the mold, which would break the face of the mold, causing scabbing or lumps on the face of the casting. In some cases where there is extra danger of scabs being produced, the reliability of the bed may be increased by first venting it with a $\frac{1}{4}$-inch vent wire, as shown in Fig. 9, before the $1\frac{1}{4}$-inch depth of facing sand is shoveled on. When this latter plan is used, the vents from the face, made with a $\frac{1}{8}$-inch vent wire, need not extend to the cinder bed.

BEDDING IN.

14. Objectionable Method.—The methods that have so far been described are the proper ones for forming the bottoms of molds that are bedded in; there are other methods in use, however, that are neither mechanical nor proper. Thus: A workman having to mold the pattern shown in Fig. 11, makes a hole in the floor and fills it with sand, without doing any packing; or else he builds a mound of soft sand. On this he lays the pattern and jumps on it; after which he digs out the sand from the lugs, again jumps on the pattern, gets off, and digs out more sand from under the lugs, and so on until the whole of the under surface of the pattern is in contact with the sand, if the pattern does not break in the meantime. Then he lays a wooden block on the pattern, and pounds it with a heavy sledge hammer, as shown in Fig. 18, until the face of the pattern is bedded

FIG. 18.

into the sand to a depth of from $\frac{1}{2}$ to 1 inch, depending on the molder's weight and strength and the weight of the sledge. This operation is repeated until the mold is made or something breaks. This is an illustration of how molding should *not* be done.

Some patterns may be bedded in by merely forcing them into soft sand; but care and good judgment are required in doing such work. When a molder uses this method indiscriminately for all bedded work, the practice is certainly reprehensible.

15. Ground Beneath Prepared Beds and Molds.
Having dealt with the preparation of soft or hard beds to
form the bottom of molds, attention is now called to the
necessity of knowing, as far as possible, the condition of the
earth or ground below the bed.   Many castings have been
lost because the ground underneath the beds of the molds
was in a soft condition.   The prepared bottom of the molds
may be made as hard as possible and the casting be badly
swollen·or distorted or all the metal lost out of the mold if
the under ground is not solid.   If there is any doubt as to
the solidity of the under ground, the only reliable procedure
is to dig down and test the under ground with pick and
shovel, or a fair idea of the condition of the ground may be
obtained by driving down pointed iron bars.   Where heavy
or deep castings are to be made over untested ground, it
may often be necessary to dig down from 2 to 4 feet below
the level of the bottom of the molds, in order to be certain
that the earth is sufficiently solid to withstand the pressure
of metal that will be put on the mold when it is poured.   A
source of trouble in many foundries arises from the careless
manner in which holes left by old molds have been filled up
with sand.   Some molders, after taking out a deep casting
that has been made in the floor, will merely throw in some
water and loose sand until the hole is filled to the level of
the floor, and then in a few days, dig out the same hole to
make a shallow casting, without going down to solid ground;
many castings have been badly strained or lost on this
account.   When deep castings (or even shallow ones) are
taken out of the floor, all the dry and loose sand should be
shoveled out and then dampened and tempered before being
shoveled back into the hole.   This sand should be shoveled
in in courses of from 5 to 8 inches deep, and butt rammed
until solid.   If this practice is followed, any molder who
may desire subsequently to make other castings over or near
the same place can rest assured that the ground under or
at the sides of his mold will be solid, and if his mold also is
firmly made, he can produce a casting free from swells or
strains on its lower side.

16. Molds With Bottom Projecting Cores.—There are many molds having bodies of sand extending upwards from the bottom that are not covered with iron until the metal has nearly reached the top of the mold; flat-bottomed annealing pots afford an example of this. Figs. 19 and 20 illustrate this form of mold, Fig. 19 being partially broken away at a to show the projection in the mold.

The gases generated at the bottom of such molds when the pouring is commenced endeavor to rise upwards through the vents formed in the projection; this, together with the

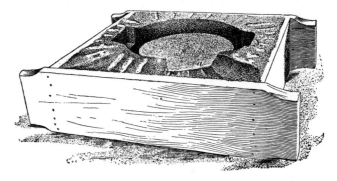

FIG. 19.

fact that the pressure of the metal over the top of the projection at a, Fig. 20, is less than at the bottom b, is liable to give trouble. Then, again, when the metal commences to cover the flat surface of such projections as a, its movement is much slower than the rise of the metal at the bottom of the mold; the result is that if there should be any blowing or bubbling of the iron as it comes to the top of the projection a, the rise of the metal will be so slow in creating pressure over the face of the projection that scabbing will take place, endangering the safety of the casting because of the liability of the iron getting into the vent holes leading down from the face at a to the cinder bed or other outlet for the vents. To prevent the projection from blowing the casting, the surface at a is made as soft as practicable and the large vents not brought nearer than within $\frac{3}{4}$ inch of the top surface of the projection. This is done by ramming up the sand firmly to within $\frac{1}{2}$ inch of the top and then striking it off with a straightedge cut away at the ends, as in Fig. 9. The bed having been struck off $\frac{3}{4}$ inch below the top of the surface, a $\frac{1}{4}$-inch vent rod is used to vent closely all over the area down to a cinder bed C or other outlet for the gases. The tops of these vents are then stopped up with the end of

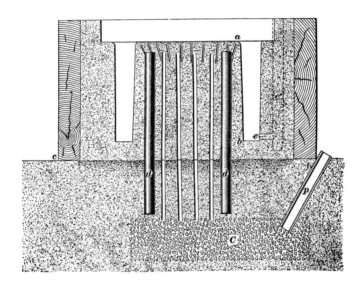

FIG. 20.

the finger, and a very open grade of facing sand shoveled in over them and pressed down with the fingers and palm of the hand as softly and evenly as can be done. This lightly packed sand should extend about $\frac{3}{8}$ inch above the level of the face at a, so that the top of the sand may be struck off to give a smooth and finished face to the projection. Before striking off this extra $\frac{3}{8}$-inch thickness of sand, the face is closely vented with a fine wire, about $\frac{1}{16}$ inch in diameter,

to a depth of about 3 inches, thus connecting the face vents
with the large $\frac{1}{4}$-inch vents, as shown in Fig. 20.   The $\frac{3}{8}$-inch
thickness of sand is struck off and the surface finished with
a trowel, leaving the top of the projection as it appears at a.
The mold is shown made above the level of the floor, but it
can be made in a hole dug in the floor, so as not to require
any part of the flask but the cope; in fact, it would be better
to make such castings in the floor, if the shop arrangement
permits digging, as then there cannot be any danger of a
run-out at the bottom of the drag at c.   This casting may
be poured by an inlet gate e connecting with the upright
gate, or sprue, e.   The cope is not shown; it is a wooden
one, 6 inches deep, with bars about 5 inches apart.   In pouring this mold, the cope is held down with about 800 pounds
of iron.   The riser should be kept closed during the pouring so as to maintain a pressure of gases in the mold, as
this will assist the gases in finding their way down to the
cinder bed.

17.   Rodding Projections.—There are a great many
projections like that shown in Figs. 19 and 20 that must be
rodded in order to make sure that the buoyancy of the
metal will not lift them.   To make sure of this in the casting shown, six $\frac{3}{8}$-inch round rods, as shown at d, are **clay-
washed** or covered with flour paste to make the sand stick
to them, and then driven at equal intervals around the
circle to the depth shown in Fig. 20.   In the hands of a good
molder this projection may stand without rodding, but it is
always advisable to take as few chances as possible.   The
question as to whether it is wise or not to rod a projection
in this manner is often one of judgment, as the condition of
the patterns, flasks, and sand often has much to do with
determining what plan it is best to follow.   All material
that is lighter than liquid metal will rise to the top and float,
just as cork will float on water.   Sand is a lighter substance
than iron, and for this reason, if the rammed sand is not
held down the iron may get under it and cause it to float.
The **specific gravity** of cast iron ranges from 6.9 to 7.4;

brass, from 7.8 to 8.4; whereas rammed molding sand ranges from 1.4 to 1.8.

NOTE.—The specific gravity of any substance is the ratio of its weight in air compared with the weight of an equal volume of pure water at 62° F. For example, 1 cubic foot of cast iron weighs about 450 pounds and 1 cubic foot of water about 62.355 pounds at 62° F. The specific gravity of cast iron is, therefore, about $\frac{450}{62.355} = 7.2$.

Taking bulk for bulk, a cubic foot of iron weighs about 450 pounds, whereas a cubic foot of rammed sand weighs about 100 pounds, the specific gravity of iron thus being about $4\frac{1}{2}$ times that of rammed molding sand. This means that a projection of rammed sand, such as that seen in Figs. 19 and 20, would have to be about $4\frac{1}{2}$ times heavier than it is before it would remain in position as molded were it not assisted by other forces. One aid to this is the adherence of the tempered and rammed sand. This prevents it from being readily separated. We find this principle illustrated in the necessity for using parting sand on sleeked joints, in order to separate the sections of molds, as previously described. This adherence in the sand of the projection would leave but little risk if the rods at *d* were omitted in this special casting, provided it was well rammed at the bottom. This is where the danger lies with all such work; any softness at the lower edge of such projections as this allows the metal to undermine the projection, and if the metal once gets underneath it, it will rise unless held down by rods or some other means. Then, again, it is necessary to guard against these projections being loosened at the bottom when jarring the pattern to draw it from the mold. Still further, patterns are often deficient in taper, so that in trying to draw them, the whole projection may be started or lifted from the bottom and so leave an opening for the metal to pass under. If any doubt exists as to the safety of a projection, it is better to use the rods *d*, and so prevent possible loss of the casting. Evidently no positive rule can be given for such work, but a study of the principles here outlined, together with his knowledge of the work, will enable the molder to arrive at a right decision in any given case.

18. Taper on Patterns.—In the last article, reference was made to the danger of **starting** projections at the base

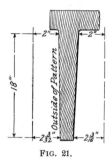

FIG. 21.

owing to a lack of taper on the patterns. As a rule, a pattern should be given all the taper that can be practically allowed. The greater the taper, the less labor will be required in finishing the mold, and, also, the longer the pattern will last. The deeper a pattern is buried in the mold, the more taper it should have. In cases where the bottom of a deep casting must be of nearly the same thickness as at the top, the pattern must not have more than a certain taper, as explained in Part 1. Where the conditions are not exacting, however, it would be much better to double the allowance of $\frac{1}{16}$ inch to the foot there given. This would give $\frac{1}{8}$ inch per foot of taper for whatever depth the patterns might be rammed in the sand. In designing work calling for such projections as are shown in Figs. 19 and 20, every effort should be made to have not less than $\frac{1}{8}$ inch per foot of taper on the inside projections. Such a pattern would, therefore, have its top and bottom dimensions as shown in Fig. 21, if the length of projection were 18 inches, as indicated. It will be noticed that $\frac{1}{16}$ inch per foot is allowed for the outside taper and $\frac{1}{8}$-inch taper for the inside, where it is most needed. Not only should patterns have good taper, but they should also be well provided with arrangements for drawing the pattern. In deep patterns, this calls for draw-plates, which are screwed to the tops of the patterns.

19. Drawing Deep Patterns.—Where a deep pattern is used to give parallel castings, or where difficulty is apprehended in getting a pattern out of the sand, it is sometimes necessary to draw it with the foundry crane. The foreman should give directions while the pattern is being drawn, in order that it may be raised evenly. If one side is drawn faster than the other, it not only causes the pattern to bind in the sand, but in the case of molds having projections, there is great liability of starting the projections from their bases, even though they be well rodded. Starting the base might permit the iron to pass in at the line of separation and so into the vents, even though it did not lift the whole projection. To have the iron run into the vents at the base of the mold is as bad as lifting the whole projection, for in either case the casting will be lost.

CLAMPING AND WEIGHTING THE MOLD.

BUOYANCY.

20. Lifting Pressure of Molten Metal.—When the molds have been closed, it is necessary either to clamp or weight down the copes, to resist the lifting force of the metal. This is readily understood when we consider that a liquid will support a body having a smaller specific gravity than itself. Sand is lighter than iron, and for this reason it will float on the surface of that metal, unless it is held down. The actual force required to hold rammed sand down depends on the height of the column of molten metal (that is, the head-pressure) and the weight of the core or cope that is liable to be floated by the metal; the cope or core, like a ship floating on water, will sink until it has displaced liquid iron equal to its own weight, when it will float. To make this subject as plain as possible, we will suppose that A in Fig. 22 is a tank instead of a mold, and has been filled with water to its top, as shown in Fig. 22 (*a*). If the block of wood B is placed in this tank, as shown in Fig. 22 (*b*), the water will run over its sides until the block comes to rest. If the water that ran over the sides were collected and weighed, it would be found to equal the weight of the block *B*. In other words, this block, in coming to rest, sank to such a depth as displaced a body of water equal to its

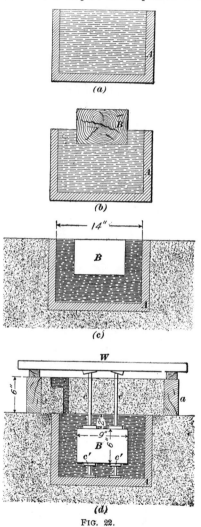

own weight. If we desired to sink this block *B* to the level of the top of the tank, as shown in (*c*), sufficient weight would be required on the block to equal the additional weight of the water that would be displaced by the block sinking to the depth shown. Carrying this illustration still further, it might be desired to submerge the block, as in Fig. 22 (*d*). To do this, but very little additional weight would be required, because as soon as the block was just covered, the weight of the water above it would be practically the same as the extra amount displaced below the block. This experiment illustrates the principle involved in w e i g h t i n g down cores or copes, the principle being the same both for water and for liquid metal. The only difference between the two is that if the block *B* were a core submerged in liquid iron, more weight would be required to hold it down, because iron has a greater specific

Fig. 22.

gravity (or, is heavier) than water. A cubic foot of pure water at a temperature of 60° F. weighs about 62⅓ pounds, and a cubic foot of ordinary gray cast iron about 450 pounds.

21. Submerged Cores. — Suppose the liquid to be molten iron instead of water, and that the cope a, Fig. 22 (d), is placed on top of the mold; find the weight W that would, by means of chaplets c, keep the core B from rising. The weight is computed in the following manner: Assuming the core to be 24 inches long, 9 inches wide, and 6 inches deep (and its ends to be free), its volume is $24 \times 9 \times 6 = 1,296$ cubic inches. A cubic inch of cast iron weighs about .26 pound; hence, the weight of the iron displaced by the core is $1,296 \times .26 = 336.96$ pounds. The weight of the core itself is $1,296 \times .06 = 77.76$ pounds, .06 being the weight of a cubic inch of rammed sand. The buoyancy of the core is therefore the difference between these two amounts, or 259.2 pounds, which is the weight required to keep the core from rising.

If the core were supported by prints instead of being free at its end, its additional length and the sand over the prints required to stop them up would also have to be deducted from the 259.2 pounds; but as the core in the present case is supposed to be held up by the chaplets c, we have here only considered the length that is submerged. Having found the weight necessary to hold down the core, the next step is to find the weight of the cope.

22. The Cope. — Assuming the cope to measure 34 inches by 24 inches by 6 inches, its volume is 4,896 cubic inches, which multiplied by .06 gives 293.76 pounds as the weight of the cope. This is for wooden copes and is accurate enough; if it were iron, we should have to find the weight of the sides and bars separately, and then that of the sand, and add them together. Having now found the weight required to hold down the submerged core, and also the weight of the cope, the additional weight necessary to hold down the cope in connection with the core must now be computed. To do this, first find the area of the cope's casting surface; this is 24 inches × 14 inches = 336 square inches. This area is multiplied by 6, the height in inches of the head or gate, giving 336 square inches × 6 inches = 2,016 cubic inches. This result is multiplied by .26, which gives 2,016 × .26 = 524.16 pounds, the fluid pressure tending to force up the cope. Deducting from this weight the weight of the cope, gives 230.4 pounds as the weight necessary to hold down the cope alone. Adding to this the weight necessary to hold down the core, we have 230.4 + 259.2 = 489.6 pounds, the total weight to be placed on the cope in Fig. 22 (d).

23. Cores Partially Submerged.—There are cases where the cores are only partially submerged, their upper surfaces being in contact with the cope, as in Fig. 22 (c). In calculating the pressure on such a partly submerged core, we must compute the area of the lower surface of the core and also the area of that portion of the cope that has metal beneath it. Each of these is then multiplied by the distance to the top of the highest point to which the metal may rise in pouring the mold.

In the case of a core submerged as in view (c), and intended to be covered with a cope, as in view (d), the computation is as follows: The lower surface of the core has an area of 24 × 9 = 216 square inches. Multiplying this by 12, the height in inches from the bottom of the core to the top of the pouring basin, we have 2,592 cubic inches, which multiplied by .26 equals 673.92 pounds, which is the upward pressure on the core. From (c) we find that the width of the cope having metal in contact with it is 14 − 9 = 5 inches; the portion of the cope surface in contact with the metal is therefore 24 × 5 = 120 square inches, which multiplied by 6, the height of the cope, equals 720 cubic inches, and this multiplied by .26 gives 187.2 pounds as the upward pressure on the cope. This added to 673.92 pounds gives 861.12 pounds as the total upward pressure. Deducting from this the weight of the core and cope leaves 861.12 − 371.52, or 489.6 pounds as the weight to be placed on the cope.

METHODS OF COMPUTING WEIGHTS.

24.   Chart Method.—The weight required to hold down the cope and core may also be ascertained by the method shown in Fig. 23.   Here we merely draw an outline

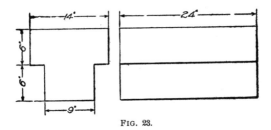

FIG. 23.

or chart of the form and sizes of the lifting surfaces, together with the height of the fluid head.   This done, we compute the cubical contents of the form thus obtained and multiply it by the decimal .26.   The volume of such a block as that in Fig. 23 is $24 \times (14 \times 6 + 9 \times 6) = 3,312$ cubic inches, which multiplied by .26 gives a weight of 861.12 pounds. Deducting 371.52 pounds (the weight of the core and cope), we have 489.6 pounds as the weight to place on the cope, the same result as by the former method.   This method of drawing a chart block of the lifting surfaces and calculating therefrom the weight to put on the cope is a very convenient one and is the one that is adopted when computations are made from drawings.

25.   General Rules.—To find the weight necessary to hold down submerged cores, first compute the cubical contents of the space occupied by the core and multiply it by .26; then deduct from this the weight of the core.   In other words, the lifting or static pressure on a submerged core is the number of pounds of iron it displaces minus the weight of the core.

To find the weight in pounds required to hold down a cope, multiply the lifting surface of the cope by the height of the head above this surface and the product by .26.

Where one wishes to compute the weight approximately, let him imagine a weight having a face like the lifting surface, and its sides extended up to the top of the pouring gate, according to the scheme shown in Fig. 23.

To find the pressure on the sides of the mold, multiply the vertical height of a side, measured from the top of the pouring gate to the center of gravity of the side, by the decimal .26, and the result will be the pressure in pounds per square inch on that side.

To find the pressure on the bottom of a mold, multiply the bottom area covered with metal by the vertical height to the top of the pouring gate and by the decimal .26, which gives the pressure in pounds.

26. Extra Weight Required on Cope.—While it is true that the foregoing rules for weighting down copes, etc. will give just the weights required under the simplest conditions, there are other conditions affecting the results that must be considered and that will often demand more weight than that given by the rules. This is due to the fact that there is an instant when the metal comes up suddenly against the lifting surface, during which a sudden pressure is exerted that is greater than that due to the height of the head, the latter being merely the steady pressure that will be exerted by the liquid when at rest. When pouring a mold, it generally takes from 10 to 50 seconds (sometimes more) to fill it with metal, whereas when the mold itself is filled, the pouring gate may fill in less than a second, thereby obtaining a head-pressure in a moment's time that, owing to the suddenness of its creation, may in some cases be so great as to call for one-fourth to one-third more weight than the static-head pressure obtained by the rules just given. The higher the top of the pouring gate is above the cope's lifting surface, the greater will be this extra pressure.

Then, again, some molds will be poured with more than one ladle, and the more ladles that are used, the greater will be the pressure; this is due to the increased pressure created by the metal as it flows from the ladle directly into a

.gate, as shown at *e*, Fig. 24. This increased pressure may be equivalent to the pressure of a head of one-fourth to one-third the height of the ladle's lips from the top of the gate.

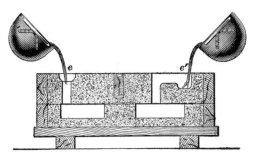

FIG. 24.

If the pattern is gated and poured as shown at *e'*, less weight will be required to hold down the cope.

27. Momentum Lift.—In addition to the weight rendered necessary by the head-pressure, extra weight is required to allow for the **momentum lift** caused by the sudden stopping of the inflowing iron at the moment the mold is filled. The amount of this depends, briefly, on the character of the pouring system, the speed of pouring, the number of ladles, and the square inches of lifting area that the metal will suddenly rise up against, as well as the height of the pouring gate or flow-off risers above the face of the cope's lifting surface. Enough has been said to demonstrate the wisdom, and often the necessity, of placing more weight on a cope than is called for by the head-pressure, and the molder must exercise good judgment in this matter.

28. Effect of Dull Iron on Buoyancy.—The lifting force of the molten metal depends in a measure on whether it is hot or dull. If the metal is dull, in most cases it will exert less pressure than if it were hotter and therefore more fluid. On the other hand, the duller the iron is, the more apt it is, in molds having risers or flow-off gates, to have its pressure approach that due to the pouring basin's height, which is generally higher than that of the top of the risers or of the flow-off gates. Often the metal will *freeze* at the entrance to the risers, or it may come up the risers so sluggishly as to retard the flow of metal out of them and so cause the head-pressure to approach that due to the height of the pouring basin. In the case of thin castings, if the metal is dull enough to freeze in the risers, it is not very apt to exert a great lifting pressure on the mold. If, with thick castings, the risers or flow-off gates should freeze up or flow sluggishly, there will be exerted a lifting pressure due to the full height of the pouring basin's head.

29. Computing the Static, or Head, Pressure.— Some molders compute the **head-pressure** on the cope by taking the height from the top of the riser or flow-off gate, the top of which is often located 4 to 6 inches below the level of the top of the pouring basins or gates. This is rarely a safe practice, as risers or flow-off gates may solidify or be blocked up so that the metal cannot flow freely through them. If it were always possible to count on having hot iron and enough room in the risers or flow-off gates to carry off the metal as fast as it could be poured into the mold, the height of risers would then, as a general thing, determine the pressure. Nevertheless, the safe plan is to figure from the highest point it is possible for the metal to reach in the pouring gates or risers, and then allow extra weight on the cope.

30. Weights for Holding Down Copes.— In weighting down molds, many founders use pig iron piled in separate pigs on the cope, or else place the pigs in stout wrought-iron rings, to be hoisted into position by a crane. Others, improving on this, cast bars ranging from 1,000 to 2,000 pounds in weight and from 3 to 6 feet in length, with hooks cast in them for convenience in handling with a crane. Other foundries preserve bad castings or take lumps of heavy scrap iron for flask weights, and handle them in the best way they can. When flask weights are required,

it pays in the end to have them as handy in f o r m and size as possible; this r e f e r s t o weights for medium and large castings. For small castings, light weights f o r s n a p flasks, etc., are required; these latter are generally m a d e a b o u t $1\frac{1}{2}$ inches thick and of a size to cover the entire surface of the cope, if they are not burdensome for one man to handle. These weights generally h a v e holes in their centers and outer corners for pouring through, as s h o w n i n Fig. 25, which shows a section of a weight w and a snap-flask mold. The under surface of these

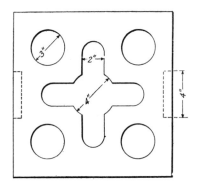

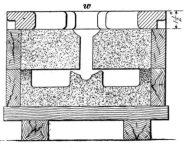

FIG. 25.

weights should be as smooth and true as they can be cast. A snap flask rarely requires more than one such weight.

CLAMPS FOR FLASKS.

31. Types of Clamps and Their Use.—Shops that do much roll-over work should have a number of **clamps** that are adapted to their needs. These clamps are made both of cast and of wrought iron and should be as handy in size and form as conditions will permit. Clamps of the forms generally used in rolling over the drags and in holding flasks together when a mold is being poured are shown in Figs. 26 to 28. Many patents have been taken out for improvements in clamps, the chief features being that their length can be changed or they can be used without wedges. In clamping a flask preparatory to casting, it is not safe to drive in wedges with a hammer, as this may jar the cope and cause the sand to drop. As a rule, flasks should be clamped by means of a clamping iron, as shown at *c, c′*,

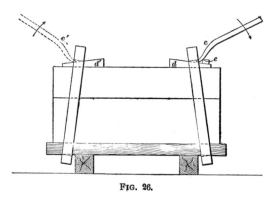

FIG. 26.

Fig. 26. The clamping irons are made with wedge-shaped points, so as to enter the small openings between the cope and clamp, and to give good leverage in either direction, as shown by the arrows. These clamping irons are, as a rule, made of old files, the points of which are turned up and sharpened.

In Fig. 27 are shown a wrought-iron clamp (*a*) and a cast-iron clamp (*b*), such as are commonly used for clamping flanged flasks, as shown in Fig. 28. Such flasks are generally used for dry-sand molding. Cast-iron clamps are usually tapered on the inside to permit of their being molded, the taper being shown, somewhat exaggerated, at *t*, Fig. 27 (*b*). In placing clamps on

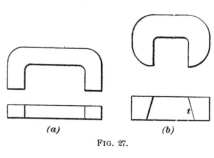

(*a*)   (*b*)

FIG. 27.

what exaggerated, at *t*, Fig. 27 (*b*). In placing clamps on the flanges, they should be set so that the wedge will enter them at their largest side, as at c, Fig. 28, which shows

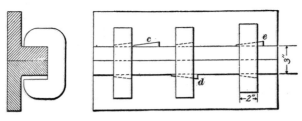

FIG. 28.

the wedge about to enter the space between the clamp and the flange of the iron flask; when driven, it should appear as shown at d.

Molders frequently try to clamp flasks by driving wedges at the top of the clamps, as shown at e. This is wrong; for by trying to drive a wedge in at the top, the weight of the clamp must be lifted, and in doing this a constant jarring takes place, making it difficult to get the clamp to a solid bearing. Not only is it difficult to tighten a clamp by top wedging, but it also requires much more time.

32. Strength of Clamp.—Where the work is such that the clamps must resist a heavy pressure, as in the case of casting rolls, pipes, etc. on their ends in dry sand, the wrought-iron clamp should be given the preference, as cast-iron clamps are not to be relied on in such work. Owing to the breaking of cast-iron clamps, castings have often been lost and men have been burned.

The proper thickness for clamps can be obtained by figuring the pressure on the flasks to be held together, and then deciding the distance apart that the clamps will be set along the flange of the flask. Knowing this, and allowing a stress in the clamp of 15,000 pounds per square inch for good wrought iron and 5,000 pounds per square inch for good cast iron, one can form a very good idea as to the proper thickness for clamps.

33. Floors for Holding Down Copes.—Many foundries that handle a standard class of work that can be bedded in have a large part of their floor area dug to a depth of from 3 to 6 feet, according to their requirements, and then place iron beams or binders in the bottom of the pit, about 3 feet apart. A plank flooring is formed over the top of these binders; from the ends of the binders wrought-iron straps or bolts are run up to the level of the floor. This pit is then solidly rammed with sand to the level of the floor, after which the holes are dug out for molding the patterns. The molds are made after the usual manner, and binders are placed across the top of the cope directly over those in the bottom of the pit; after which, bolts are extended from the top binders to connect with those extending from the bottom ones, and the cope is bolted down in such a manner that the bolts will have to be broken before the cope can be raised by any pressure that might come on it. The practice of using these bolting-down floors instead of weights is a good one.

GREEN-SAND MOLDING.

(PART 3.)

IRON MOLDING IN GREEN SAND.

DETAILS OF THE MOLD.

JOINTS.

1. Joints for Parting Circular Forms.—The simplest form of **joint** and the one most used is the straight joint, the manner of making which was described in *Green-Sand Molding*, Part 1. Joints for parting circular forms are shown in Figs. 1 and 2. When the pattern is divided as shown in Fig. 1, it is customary to make the mold with half the pattern in the drag and half in the cope. The parting line of the pattern and the parting line of the mold are made to lie in the same plane, so that in lifting the cope, half the pattern is lifted with it and the other half remains in the drag. If the solid pattern shown in Fig. 2 is to be used with a straight joint in the mold, it is necessary to bed half the pattern in the cope temporarily. The drag is then put on and rammed up, the mold rolled over, and the cope shaken out, placed on the drag, and rammed up again. The solid pattern can also be used in the bedding in process as explained in *Green-Sand Molding*, Part 2, and the sand is smoothed down to the level of the middle of the pattern. A cope can then be placed over it and rammed up in the usual manner.

§ 42

While the method of dividing the pattern as shown in Fig. 1 can be followed to advantage in many cases, there are others where it is better to make the pattern of one

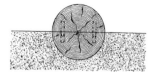

FIG. 1.

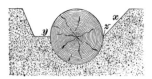

FIG. 2.

piece. Another method of molding when a solid pattern is used is to place the pattern on a follow board and ram up the drag in the ordinary way. The mold is then turned over and the joint cut down to one of the two forms shown at x and y, Fig. 2.

The joint shown on the right at x is bad practice, as some sand will often be left sticking in the sharp angular pocket formed against the pattern at z. When sand does stick in such pockets, it is difficult to patch them without causing a heavy " fin " on the casting or taking chances of a " crush " at the joint. Then, again, such a form gives a very poor bearing for gaggers, the setting of which is explained later.

By making a joint as at y, on the left of Fig. 2, every opportunity is afforded for a good bearing for gaggers and for obtaining a clean lift. Should any of the joints break and require patching, this can be done without much danger of leaving large fins on the casting or causing the mold to crush, owing to the flat surface at the joint, which gives a good guide for patching any broken edges. It is chiefly in heavy castings that cutting down in this manner is practiced.

These two methods of parting the mold when using a solid pattern are rarely employed except when the cope is shallow or it is desired to avoid cutting the bars of the flask to conform to the shape of the pattern. Small patterns are usually divided as in Fig. 1 or provided with follower boards.

2. Joints for Irregular Forms.—Fig. 3 illustrates a method of making joints by having the parting line of the cope and nowel conform to the shape of the joint of the pattern. The illustration represents the end view of the cope a, drag b, and bottom board c, the cope being in place for ramming up. The dotted line at d shows the lower line of

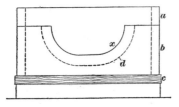
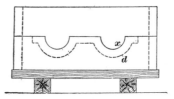

FIG. 3.　　　　　　　　　　　FIG. 4.

the pattern, while x is the face line of the pattern and the line of separation between the cope and nowel. Another example of this is seen in Fig. 4, which shows the side view of a cope and drag, the lines of the pattern being at d and x, as in Fig. 3.

3. Three-Part Molds in Three-Part Flasks.— Many patterns are of such a form as to require two or more parting lines. Sheave wheels and wheels having flanges are the most common representatives of this class of castings. The most common way of casting such pieces is to have as many parts to the flask as there are parts in the mold. One method of molding and casting a flanged pulley is illustrated in Figs. 5 to 7. The part of the pattern p is placed

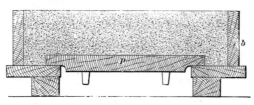

FIG. 5.

on the follow board and the drag b is placed around it. This part is rammed up and struck off as shown in Fig. 5. The bottom board is then put on, the mold turned over, the pattern section q put on, parting sand sprinkled on the mold, and the cheek or intermediate part of the flask f put in place. This part of the mold is then filled up with sand, rammed, and struck off, as shown in Fig. 6. Parting sand is then sprinkled over the joint, the cope put on, the sprue pin

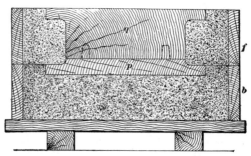

FIG. 6.

located, and the cope rammed. The sprue pin is then removed and the pouring basin e, Fig. 7, formed. The cope a is lifted off and the pattern q withdrawn from the cheek. The cheek f is then lifted and the pattern p withdrawn

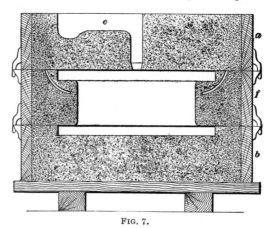

FIG. 7.

from the drag. The drag is finished, the cheek put in place, the groove of the pulley vented, the cheek and cope finished, and the cope placed in its position, as shown in Fig. 7. The mold is then complete and ready for the metal to be poured.

4. Making Three-Part Molds in Two-Part Flasks.
A three-part mold may be made in a two-part flask by having
an intermediate body of sand between the cope and the

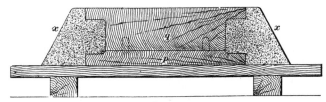

FIG. 8.

drag. This is illustrated in Figs. 8 to 10, in which the

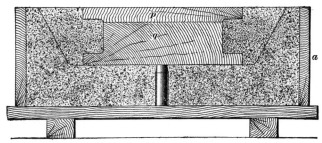

FIG. 9.

pattern described in Art. **3** is used. The pattern is placed

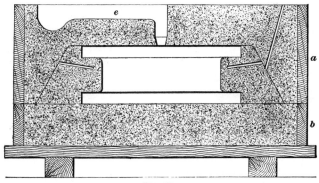

FIG. 10.

on a mold board and weighted down, after which sand is
packed around it to form a joint, as shown at *x*, Fig. 8.

52—30

The cope a is now set on and rammed up, after which the whole is turned over, when it will appear as shown in Fig. 9. Next, the drag b is set on and rammed up, after which it is lifted off and the pattern section p drawn; the drag is then set back, and both parts clamped and rolled over. The cope is next lifted off and the pattern section q drawn; the mold is then finished and the cope set back in place, when the whole will appear as shown in Fig. 10. Parting sand must be used on all the joints.

In venting the cope, it is only necessary to run vents from the parting line with the intermediate section to the top of the cope. The gases will find their way along this parting line and out through the vent.

5. Starting the Joint in Lifting.—It is important not only that the joint in the mold be properly constructed,

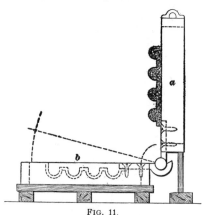

FIG. 11.

but also that means be provided for properly separating the parts, as a good lift often depends on the manner in which the cope is started. It is necessary to start some copes evenly all over the joint, while others part better by starting one side before the other. In light work, there are many copes that are best raised by being rolled up, as in Fig. 11. In rolling up a cope, it may be advisable to have it go upwards and toward the hinge, as shown by the dotted lines in Figs. 11 and 12; or on the other hand, it may go upwards, first describing an arc away from the hinge and then toward it, as shown by the dotted line in Fig. 13. This movement can be given by arranging the hinges as shown in Figs. 12 and 13, from which it is clear that the matter of having a cope go upwards

from or toward the hinged side can be controlled, thereby
assisting in getting good lifts when a movement in either

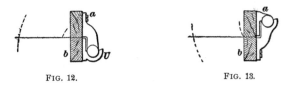

FIG. 12.   FIG. 13.

direction will do this.   Of course, the farther from the joint
the center of the hinge is, the more rapid will be the outward
or inward movement.

FOLLOW BOARDS.

6.   Their Use in Forming Joints.—In making the
mold for small castings (either of iron or brass) that have
irregular joints, much time and labor may be saved by using
a **follow board** that will form the joint; then, when the
board is lifted off, parting sand can be sprinkled on and
the joint is ready for the cope.   Many persons making light
work have the follow board so perfected that it is often difficult to perceive where the joint is on the casting.   There
are four classes of follow boards: (1) the *wooden* follow board,
which is carved out to give the desired shapes to the joints;
(2) the *sand-and-composition* follow board; (3) the *plaster-of-
Paris* follow board; and (4) what is called the *match board* or
match plate.   The sand-and-composition follow boards are
the ones generally used.   Follow boards are sometimes called
odd sides or *matches*.

7.   Sand Follow Boards, or Matches.—The usual
method of making sand follow boards is to ram up a drag
(harder than usual), and then, after making a good firm
joint, set upon it a frame having nails driven in at its sides,
as shown at *n*, Fig. 14; these nails are driven into the frame
to hold the sand when using the match.   When this frame is
set on, new molding sand (tempered with very thick clay

wash or else made up of 1 part flour to 10 parts sand) is
shoveled in and rammed as hard as possible, after which a
bottom board like the one shown at *c* is nailed on to the
bottom of the frame and the frame and board then lifted off.

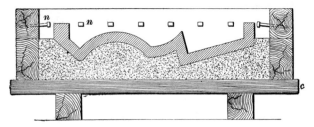

FIG. 14.

The surface of the match is now sleeked, to repair any
broken edges or parts that may have stuck down in lifting
it off, and a little molasses water is blown over the surface
to strengthen the joint.   The molasses water used in finishing the joint may be blown from the mouth or from a special

FIG. 15.

sprinkler attached to a pair of bellows, as shown in Fig. 15.
Another form of sprinkling device, known as a blow pot, is

FIG. 16.

shown in Fig. 16.   Fine sprinkling
may be done by tilting the pot and
blowing with the mouth so that
the current of air strikes the liquid
as it escapes.   When it has been
sprinkled, the match should be
set aside for a day or two to
harden, after which it is ready for use.   In light work it is
of the utmost importance to preserve the joint edges of

sand matches, so as to keep them sharp and unbroken. It will help to preserve them if all the sharp edges are stiffened with nails driven close together, allowing the heads of the nails to come flush with the face of the sand forming the match; even then, the edges may become ragged and cause bad joints. A match made from molding sand, or what is termed a green-sand match, is ready for use as soon as it is made. Most molders give these matches a slight sprinkling every day after the day's work is finished, for they do not work well when in a dry condition.

8. Composition Follow Boards.—In place of ramming the frame, Fig. 14, with molding sand, a composition may be used that will become harder and last longer than the sand match. A composition often used is one made up of fine sand, boiled linseed oil, and litharge. The sand should be very dry. Add 1 part of litharge to about 20 parts of sand, mix thoroughly, and then sift the whole through a fine sieve. Temper this mixture with the oil to the same temper as sand intended for ordinary green-sand molding. The mixture is rammed, as one would ram a mold, to a degree of hardness equal to that generally required in copes. After the ramming has been done, the bottom board is screwed to the frame. The match and the drag on which it is made are then rolled over together, the drag carefully lifted away, and the joint finally finished. Before these matches are dry, they are about as fragile as so much dry sand, and require the utmost care in handling, as well as in removing the pattern for the first time. When the match is dry its surface should be given a coating of shellac, which will prevent the sand from adhering to the surface. Before putting the match away, its edges and surface should be finished in the same manner as sand matches, using linseed oil instead of the molasses.

Molding sand should not be used for these matches, as it makes them weak; but some fine-grained sand can be used, and almost any sand of fine grain will do. If at any time the corners or edges are found to be broken, they can be mended by patching with beeswax. To form the separation between this match mixture and the sand on which it is rammed, a regular parting sand is used. For very fine work a material known as lycopodium is used. Where the match is too large to lift off the drag, they can both be rolled over and the drag lifted from the match, and the sand then carefully removed from the face of the match.

9. Plaster-of-Paris Matches. — Plaster-of-Paris matches are often used where, from the crookedness of the pattern, other classes of matches cannot be made as cheaply or as perfectly fitted or kept as true during use. This material gives very hard matches on which to ram, but great

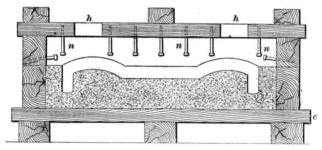

FIG. 17.

care must be taken not to break any of its edges, as, even with care, the working in and out of the pattern is very liable in a short time to cause the edges to become ragged and broken; and no durable method of patching such broken edges has yet been devised. Fig. 17 shows the match in

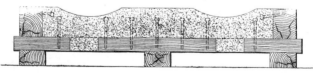

FIG. 18.

process of construction, while Fig. 18 shows it ready for use. In starting to make such a match, the pattern is rammed up in a drag and the joint made as in molding ordinary castings.

The joint should be carefully made so as to give it the best possible form, one that will give clean lifts and assist in obtaining finless and true-jointed castings. The patterns are treated with a good coat of oil to prevent the plaster from sticking to them. A wooden frame having a bottom board screwed on is then placed as in Fig. 17; both this frame and the bottom board should have plenty of nails n driven in them. In this bottom board are two holes h for the purpose of pouring in the plaster. Before pouring a plaster match, the outside of the joints should be carefully stopped up with clay, or else firmly banked up with sand, to prevent leakage. The plaster having been poured in, it is allowed to set until hard; then the drag and match are rolled over together, opened, and the sand removed from the face of the plaster with brush and water. After the face of the board is finished up smooth and the plaster is dry, it is given a coat of shellac varnish containing lampblack, and when this is dry the board is ready for use. Plaster of Paris is made by heating powdered gypsum, which consists of sulphate of lime and water. The heat drives the water out of the gypsum, leaving a powder that, when mixed with its own bulk of water, forms a creamy paste that becomes solid almost immediately.

In using plaster of Paris, the fluidity of the mixture should be regulated by the thickness of the body required. For thin bodies, 2 parts of water to 1 of plaster makes a good proportion, but for general work, 1 part of plaster to 1 part of water will be about right. The pouring holes should be as large as practicable, for in filling thin places or corners, the quicker the match is poured, the better. If a mold has any considerable body, it will shrink so much as to require being filled up with more plaster after it is poured. Before starting to pour a mold, therefore, there should be plenty of water and plaster at hand, to avoid any delay after the pouring has begun. With practice, one can estimate very nearly the amount of mixture required to fill a mold; it should be all mixed before starting to pour, especially in the case of light molds. For thick bodies one may partially fill

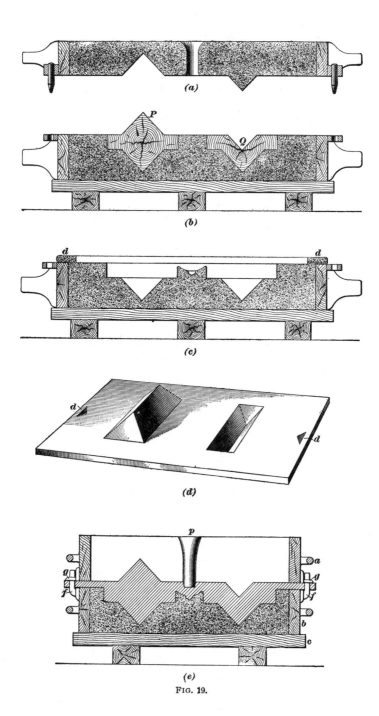

FIG. 19.

a mold and then complete the job by a second pouring; but generally speaking, plaster of Paris requires prompt handling.

10. Match Plates or Match Boards.—There are cases where **match plates,** as they are termed, will be found of value in expediting the making of joints; their construction is very simple. Fig. 19 shows the mode of constructing a match plate for two patterns, one of which comes wholly below the joint line and the other partly above and partly below it. In Fig. 19 (*b*) the drag is shown rammed up and the joint made, *P* and *Q* being the patterns. The cope when rammed up appears as shown in Fig. 19 (*a*). The manipulation so far is the same as that required for making a casting from each of the patterns. The next step is to mold the plate portion. This is done by banking sand against wooden strips from $\frac{3}{16}$ inch to $\frac{1}{4}$ inch thick. In this way the body of sand *d*, Fig. 19 (*c*), is formed. The thickness of this body of sand should be so chosen that it will make the plate strong enough. The gates are cut as though the casting was to be poured through them. The cope is then closed and the mold poured, the casting being the match plate shown in Fig. 19 (*d*). Fig. 19 (*e*) shows the drag rammed up, the cope set on, the gate pin *p* in place, and the cope ready to be rammed up. The ends of this match plate extend beyond the flask and contain holes for the flask pins to fit into, so that the mold may come together properly when it is closed. These holes are made by drilling and filing to fit the dowel-pins on the drag. This will be better understood by reference to view (*e*). The drag *b* has pins *f, f*, that are long enough to fit into the holes *d, d* in the match plate and also holes *g, g* in the lugs of the cope. This arrangement of pins and holes acts as a guide in setting both the cope and the match plate. Should there be any overlapping of joints in the castings produced, the trouble will generally be due to shaky or untrue pins. In making the match plate, as well as in using it, the pins on the flask must be carefully looked after, or properly jointed castings will not be obtained.

Wooden matches made by the patternmaker are also used. The match plate or board is of practical use only for castings that have plain outlines and are without sharp corners, cores, or projections.

GAGGERS AND SOLDIERS.

11. Use of Gaggers and Soldiers.—**Gaggers** and **soldiers,** which are described more fully in Arts. **12** and **13,** are appliances used in combination with flasks and cross-bars to enable the molder to lift and suspend bodies of sand. This will be better understood by reference to Fig. 20, which represents a cope 16 inches square by 5 inches deep. If this cope were rammed full of good and properly tempered sand, having

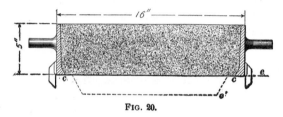

FIG. 20.

its joint level with the bottom edge of the cope at e, the sand would lift and stay suspended. Instead of the joint being level with e, it may be desirable to have the cope sand project down into the drag, as shown by the dotted line at e'. Where the sand extends more than $\frac{3}{4}$ inch below the level at e, it might not lift with the cope, or if it did, it could not be safely suspended without the use of gaggers or soldiers.

The volume of sand that can be carried without special securing varies with the condition of the sand. A coarse sand will not hold as well as a fine sand. A body of sand 16 inches square and level with the lower edge of the cope, as at e, is about as large a body as can be suspended without the use of cross-bars. Even with a body 16 inches square, it is sometimes necessary to have grooves along the sides of the flask or else projections like c, Fig. 20, as without one or the other of these the sand will be liable to slide out of the cope. While 16 inches square is given as being the largest area of sand that can be safely suspended, even that area cannot be lifted in all cases.

12. Making Gaggers.—Gaggers are made of cast or wrought iron. Fig. 21 shows the form generally used. They can be made of either square or round iron, and are usually about 4 inches long at the toe *m*, with the shank *n* from 5 inches to 20 inches or more in length, according to requirements, and from $\frac{3}{8}$ inch to $\frac{1}{2}$ inch in diameter or square.

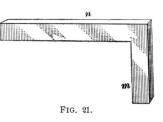

FIG. 21.

In some shops wrought-iron gaggers are used almost exclusively; while in others, cast-iron ones have the preference, as they will not spring, are cheaper to make, and have the advantage of being readily broken off to any desired length when shorter ones cannot be found.

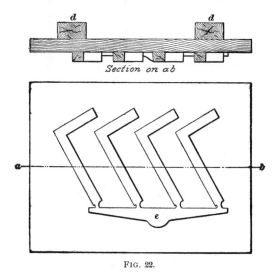

Section on *ab*

FIG. 22.

Wrought-iron gaggers are useful in work where the toe must be bent to suit slanting surfaces and joints. In some foundries objection is made to breaking cast-iron gaggers,

and to avoid breaking them, they are left sticking out of the
cope.   Gaggers sticking up in this way are liable to be hit

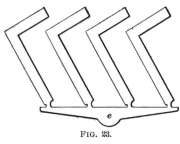

accidentally after the cope is
closed, and this may result
in the loss of the casting.
It is bad practice to leave
gaggers standing above the
surface of the cope.   Cast-
iron gaggers can be made
to good advantage in open
sand by having from four

FIG. 23.

to twelve patterns on a board, as shown in Fig. 22, and
pressing the board into a level bed of soft sand by pounding

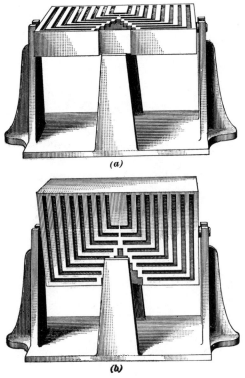

(a)

(b)

FIG. 24.

on the battens *d* with a light sledge. When the gaggers are cast, they appear as shown in Fig. 23; they are easily knocked off the runner *e* with a hammer. Wrought-iron gaggers are usually made by cutting straight bar iron into the required lengths and bending the toe *m*, Fig. 21, in a vise or over an anvil.

Cast-iron gaggers may be made very rapidly by the use of a chill mold, as shown in Fig. 24 (*a*) and (*b*). The illustrations show a gagger mold that can be used almost an unlimited number of times during the heat. It is swung on a cast-iron bedplate, supported by two trunnions that allow the mold to be turned over, as shown in Fig. 24 (*b*), which illustrates the process of turning it over for the purpose of dumping the gaggers. The metal is poured on the mold, which is then turned over, striking a stop when upside down so as to jar the gaggers loose and allow them to fall out. Both sides of the plate contain molds for gaggers, so that as soon as the plate is turned over the molder can pour the second set of molds full. This can be repeated until the mold gets hot, when it will be necessary to let it cool off for a time.

13. Making Soldiers.—Soldiers are merely strips of wood. They can be made by taking a piece of rough straight-grained board, sawed to any desired length, and cutting it into strips of any desired size. They range in size from a narrow strip to one 8 inches wide. Often these soldiers will be nailed to the sides of cross-bars, so as to assist in lifting deep bodies of sand. The soldiers, if well sustained between the cross-bars by nailing them or by ramming them firmly between the bars, will lift larger bodies of sand than if gaggers were used over the same area. In using soldiers, they must not be placed too near the surface of the casting, where there might be danger of the iron breaking the sand away from their surfaces, for if this occurred, the steam and gas from the wood would cause the mold to blow and spoil the casting. Then again, if soldiers are to remain bedded in the sand for more than a few hours, they should be well soaked

in water before being placed in position, for if they swell in
the mold, they may cause bad castings.

14. Setting Gaggers.—The main thing to be kept in
mind when setting gaggers is that, bulk for bulk, a gagger

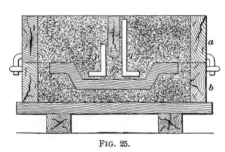

FIG. 25.

is about $4\frac{1}{2}$ times as
heavy as rammed sand.
Gaggers are used
sometimes to aid in
lifting bodies of sand
that would have a bet-
ter chance of being
lifted were the gaggers
omitted; this will be
better understood by
reference to Fig. 25, in which a body of sand about 3 inches
deep is to be lifted.   Gaggers set as at p will do more injury
than good; to be of any service they should be long enough
so that at least two-thirds of their length will be between
the cross-bars, as at q.   Then, again, where gaggers are
expected to lift a heavy body of sand, not only should they
come up well between the cross-bars or in the cope, but the
sand should be firmly peened and rammed between them.

15. Setting Cross-Bars.—In putting bars into ordinary copes, a space of from 5 inches to 6 inches between
each bar will answer for plain work; but for copes that have
bars projecting into deep recesses in the drag, or that have
them cut to permit projections to extend into the cope, different spacing and a different system of barring are often
necessary.   At s, t, and u, Fig. 26, is shown an objectionable method of setting cross-bars in the cope used for making
a long casting of the general cross-section shown at P.   One
objection is that the flat side of a cross-bar is placed parallel
with the flat face of the pattern, leaving a poorly supported,
thin, flat body of sand v.   In ramming sand in such narrow
pockets as at v, good judgment must be exercised; if the
sand is rammed too hard, the gases will not escape freely,

and scabbing or blowing is likely to result.   Another objection to the method shown in Fig. 26 is that where it is necessary to roll the cope over, the thin flat cake of sand is liable to drop off unless securely rodded, which involves having straight rods of round iron coming from the face up between the bars.   Bars used for lifting sand out of pockets or for carrying hubs or other projections should be arranged so as to have a considerable body of sand around them.

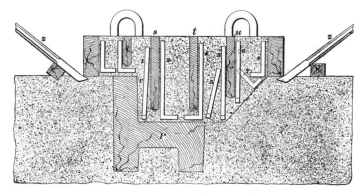

FIG. 26.

This not only lessens the dangers due to hard ramming, or lack of freedom in venting, but it gives more room for ramming up and seeing what is being done when setting gaggers, etc.   Another objection to using bars as at *s*, *t*, and *u* is that the gaggers cannot be set very readily or firmly, and the danger of a *drop-out* is greatly increased.   For work of the character here shown, the bars should be set across the mold, as indicated by the dotted outline *O O* in Fig. 27.   The position of gaggers *1*, *2*, *5*, *7*, and *9*, Figs. 26 and 27, shows lack of judgment.   The sand at *7* would be more likely to lift if the gagger were not used, as its length is only about that of the body of sand to be lifted, and iron, as already explained, is a great deal heavier than sand.   If bars could not be placed as at *O O*, Fig. 27, and it were necessary to set a gagger as at *1*, Fig. 26, it would be better to keep it about 3 inches higher and reverse the toe of gagger *2*, bringing the toe or point under the bar toward the face of the pattern. The position of gaggers *11*, *12*, *13*, and *14*, Fig. 27, in connection with the bars at *0*, *0*, represents good practice. Gagger *10* should have its toe moved very close to the face of the pattern at *v*, while *9* should be set between *10* and *11*,

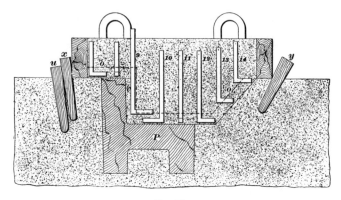

FIG. 27.

with its toe parallel to *11*. If the cope is to be rolled over, more gaggers (*13* and *14*) should be used as the height of the ramming increases. The points of gaggers against the surfaces of flat bodies of sand cannot do the harm that gaggers can when set as at *1* and *9*. The latter do not give good support and produce soft spots in the mold.

16. Driving Stakes and Starting the Joint. Fig. 27 also shows right and wrong methods for staking copes that are used over bedded-in patterns. Stakes should be driven almost parallel to the side of the cope, as shown at *x*; it is bad practice to drive them at a considerable angle, as shown at *y*; a stake driven in this manner is liable to cause poor lifts and overshot castings, because of the great angle it makes with the surface of the floor. In staking flasks for ordinary work, at least two-thirds of the length of the stake should be driven into the ground. Sometimes, to insure greater certainty in large work, it is best to drive one stake behind another, as illustrated by *u* and *x*. To assist in the lifting of such deep copes as are shown in Figs. 26 and 27, iron starting bars can sometimes be used, as shown at z, z, Fig. 26. It is important that the cope be started properly; for if it is started so as to raise one side before the other, or if it is started with a jerk, the most careful ramming and use of gaggers will be of little avail in giving a good lift. Where two or more men are required to lift the cope, it is often a good plan to first raise it an inch or two by raising each corner alternately from $\frac{1}{8}$ inch to $\frac{1}{4}$ inch and inserting a wedge to hold up the corner as it is lifted, or it may be advisable to raise one side of a cope at a time; the distance can usually be increased at each succeeding lift.

FINISHING THE MOLD.

17. Work Required After Drawing Pattern.—As a rule, all molds require more or less finishing after the pattern has been drawn, bench-work castings requiring the least of any. The majority of light-work patterns are so finely made and gated that the mold may be closed as soon as the pattern is drawn, there being no finishing whatever required; in heavy work, however, the reverse is usually true. In some cases it may take longer to finish a mold than it takes to ram it. This may be due to the intricacy of the design, or to bad work in drawing the pattern, or it may be due to the manner in which the mold was rammed up.

18. Care and Skill in Ramming.—Two molders may ram up the same pattern in the same flask, and yet one may take twice as long to finish the mold as the other. As a rule, the greater the care, skill, and time bestowed on ramming, the less time is required on finishing; the skilled and careful workman, generally speaking, so rams his mold as to require the least time in finishing.

In the case of many heavy-work molds, the insertion of a nail or rod at the corners and flanges when ramming will render them less liable to start or break when drawing the pattern; and there are but few such molds in which care in ramming will not prevent their having soft places that require to be patched. In some cases, the soft places that occur from careless ramming may be so extensive as to cause large portions of the mold to break and fall while the pattern is being drawn. In the ramming of copes especially, there is an opportunity to save subsequent labor in finishing. Some molders use so little skill and care in ramming the cope, that when it is lifted off, the sand will be soft under all the cross-bars. Where this occurs, the soft places must be pressed down solidly with the fingers, sand filled in firmly by hand, and rubbed off level with the rest of the mold by using a finishing block or straightedge before the surface is ready to be sleeked with the trowel. All this extra work can be avoided by careful ramming. When a cope is poorly rammed, the sand under the cross-bars may have to be worked over to make it solid. A cope so treated rarely gives as good and true a surface as would otherwise be the case.

19. Nails and Rods at Joints and Corners.—A judicious use of nails or rods in ramming and finishing

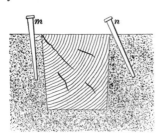

FIG. 28.

molds may prevent many castings from being defective. It is possible, however, to use them too freely or improperly. A nail should not be used as shown at *m*, Fig. 28, as when driven in this manner, it only adds weight to the edge of the mold, instead of giving it support to keep it from dropping. The proper way of using such nails or rods is shown at *n*. Here the nail is driven in such a way as to take it away from the face of the pattern into the body of the sand, where it can have a firm hold and assist in keeping the edge from dropping.

20. Patching the Mold.—Often the mold is more or less broken in drawing the pattern. Whenever practicable the mold should be mended with the hand, and then smoothed off with a finishing block or straightedge to as nearly the proper form as possible before a trowel or other finishing tool is used. Many molders patch such places with a trowel; but when one takes sand on a trowel and

FIG. 29.

presses it on to the mold, he gives the face of the patched part a smooth surface, with which the next trowel of sand will not unite as well as when the broken parts are built up by using only the hands. The objection to patching with a

FIG. 30.

trowel is that the patched part may be easily loosened and is liable to drop if the cope is slightly jarred; or it may be washed off by the friction of the inflowing metal when the mold is poured. Fig. 29 shows a molder's hand patching a broken corner with a trowel, while Fig. 30 shows the hand being used to get the part in proper form before the trowel is used. Patching of this kind may sometimes be done by placing pieces of straight boards against the sides of the mold, thus getting a perfect outline, and then pressing sand down on the mold.

21. Swabbing Broken Corners. — Many molders before starting to patch a broken mold freely wet the surface with water, thinking thereby to make the sand stick better to the broken body. Unless the broken surface is drier than the rest of the mold, it should not be moistened. If it should require moistening, however, the mouth or one of the spraying devices shown in Figs. 15 and 16 should be used; for when the water is put on with a swab or sponge, it is liable to make the surface of the broken part too wet, and this may do more harm than good.

22. Moisture in Molds.—In tempering green sand, it is given a certain degree of moisture, and when this is too great, the volume of steam created by the hot metal in the mold, during the pouring, may become dangerous. The sand will permit a certain amount of steam to pass through it without harm; but when a molder makes the under surface of his mold too wet, and then fills the mold with molten metal, the latter rapidly heats the sand to a temperature sufficiently high to change the water to steam, and this steam will liberate itself in the line of least resistance. If the wet portion of the mold has been well vented, the steam may pass off through the sand, but the chances are that the line of least resistance will be through the liquid metal. If the steam passes through the metal, it is likely to have enough force to raise (in part, at least) the body of sand that is covered by the metal. If it does this, we may expect lumps or scabs on the casting, making it defective. Or again, there may be so much steam created that it will start the mold blowing, and this may result in losing the casting; or worse still, in throwing the iron out of the mold at the risk of life and property. As a rule, it is safe to mend a broken part of the mold without first wetting it, and then after the patching is completed, to take a swab and wet the surface of the finished part, in some cases quite heavily, and still have no injurious results from steam. This is owing to the fact that in order to escape, the steam will not have to raise a body of the sand. The steam being created at the surface, the portion that does not pass off through the sand has only the iron to pass through in order to escape. In doing this, it may make the casting blow to some extent, as it would if the steam had come from the lower parts of the mold; but if the blowing is not too great from this cause, the casting will not be injured.

An illustration of what is to be expected from steam confined under the surface of the mold is shown in the fact that one can cover the surface of a body of liquid metal with water without any injurious consequences, the reason being that the steam is created on top of the iron and has simply to pass off into the atmosphere in being liberated. Let one try, however, to place the same body of water in the bottom of a ladle and then pour liquid metal in on top of the water; the result will be an explosion that will drive all the iron out of the ladle, and possibly seriously burn those near by. Dampness can only exist with safety so long as it is on the top of the metal; if it occurs underneath the metal, serious results may be expected. If the molder will bear this principle in mind when swabbing any part of his mold, he will have very little trouble with scabs or blowing as a result of an excess of moisture.

Another evil resulting from the use of too much water in finishing a mold lies in its hardening the metal at the point of extra dampness. The edges of castings can be made so hard by extra dampness in the sand at such points that a file will not cut them. Another effect of excessive dampness is to give the iron an extra amount of combined carbon at such points; this alteration of its physical nature may cause a casting to crack when cooling, or break in pieces when put into use.

23. Venting Patched and Sharp Bodies of Sand.

It is always well to vent patched parts of a mold with a

$\frac{1}{16}$-inch vent wire, for usually the patched sand will be harder and damper than the rest of the mold.   Again, in many large molds having corners, projections, etc., it is a good plan to pass a fine wire from the face downwards into the body of the mold to a depth of from 4 to 6 inches. These fine-wire vents will provide a means of escape for the gases to the larger vents, and if made about 1 inch apart over the surface most likely to scab, that evil will be avoided.

FIG. 31.

Many molders make a practice of venting almost every sharp corner or projection in large molds, and although this takes time, yet it pays in the end, for it is seldom that any delicate portion of molds so vented gives scabbed castings.   To prevent such fine venting breaking the surface of the molds, the vent wire is run through the opening between two fingers, as shown in Fig. 31. The tops of the fine vent holes are stopped up by pressing the fingers or palm of the hand over them, and then going over the holes with a little finely sifted sand, rubbing it into them with the hand; after this, the surface is neatly sleeked and then dampened lightly with a swab, or sprayed.   The success of some molders in getting large castings free from scabs, etc. is due in part to their habit of using the fine vent wire at corners when finishing their molds.

24. Using the Trowel.—Considerable skill is required in handling a trowel properly. When one first uses a trowel,

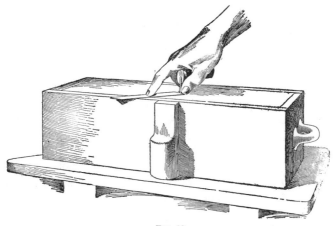

FIG. 32.

he is liable to dig into the sand and do more harm by loosening the sand than he does good by pressing it down to a

FIG. 33.

solid smooth surface. The trowel should never be kept flat on the body being sleeked, as in Fig. 32. The proper way to use it is to raise one edge slightly, as at *d*, Fig. 33; it is here, however, elevated an excessive amount for the purpose of illustrating the idea more clearly. The trowel should have its forward edge raised only about $\frac{1}{16}$ inch, this being just enough to keep it from digging into the sand and yet not leave a flat face on the sand. If one is an expert, he may, in many cases, sleek green-sand surfaces with the whole flat surface of the trowel bearing on the sand.

In handling a trowel, the first finger should project as far on the blade as convenient, so as to give a pressure to the blade, as shown in Fig. 32; a novice will usually grasp a trowel by the handle, as shown at *e* in Fig. 33.

A facing of dry blackening or silver-lead dust should rarely be sleeked on the surface of a mold with the flat of a trowel; for if the blackening does not stick to the trowel, it is liable to loosen in such a manner as to lift when the mold is being poured and cause what are called **blackening scabs** to appear on the casting.

25. Using the Sleeking Tools.—When sleeking wet blackening on cores or molds (skin-dried, dry-sand, or loam work), the trowel must be kept tilted. If at any time the flat face of the trowel or any other finishing tool touches the wet blackening, it will stick to it. Not only must the finishing tool be tilted, but it must be kept in motion, for if stationary for an instant, the wet blackening will stick to it. Considerable skill is required in sleeking wet blackening, and much experience is necessary before one can handle finishing tools in such a manner as not to start the blackening, the effect of which would be to cause blackening scabs on the casting. Some wet blackenings are so difficult to sleek that it is necessary to keep constantly dipping the tools in water, in order that they may slide more easily over the blackened surface.

26. Other Finishing Tools.—A molder should have a good set of molding tools. Some shops demand that a molder be well equipped with tools; in fact, they often go so

far as to require the molders to have tools that will fit nearly
every variation in the shapes of edges and corners that may
exist in their patterns, and if these shapes are out of the
ordinary line manufactured by regular toolmakers, they will
have the special tools made.

The trowels, lifters, and double-enders are usually made
of steel, while the other tools are often made of cast iron and
brass. Brass tools will sleek wet blackening better than
those made of iron or steel, although if steel tools are nicely
made and finished, many molders can do better work with
them than with those made of brass. Figs. 32 and 34 show
the ordinary finishing tools. These can be obtained from
dealers in different sizes. A tool box especially designed to

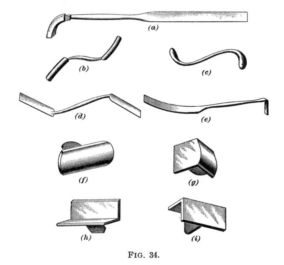

FIG. 34.

hold tools in such a manner that any one of them can be
readily found is very desirable, and the tools should always
be clean and in good order, ready for use. The names of
the tools shown in Fig. 34 are as follows: (*a*), flange and
lifter; (*b*), flute; (*c*), bead; (*d*), double square; (*e*), Yankee
No. 1; (*f*), pipe slick; (*g*), half-round corner; (*h*), inside
square corner; (*i*), square corner.

27. Sleeking and Printing Dry Blackening.

After the surface of a mold has been finished, it is often

necessary to blacken it, so that the casting will have a smooth surface and will peel better from the sand. The blackening may be dusted on or rubbed on with the hand and then sleeked down solidly with the same tools that were used in finishing the surface of the sand. For heavy castings, it is best to rub on the blackening with the hand; especially is this necessary when putting it on the sides of molds and copes that cannot be

FIG. 35.

rolled over.   In a great variety of work, the blackening can be shaken out of a cheese cloth, or other thin cotton-cloth bag, as shown in Fig. 35; or it can be scattered by the hand in the same way that parting sand is spread.   After a mold has been coated with dry blackening, it should be sleeked as soon as it can be conveniently done.   If there is any delay, the blackening is likely to absorb the moisture from the sand; with some blackenings it will then be difficult to sleek them without their sticking to the trowel, the result of which will be a badly finished mold that may cause blackening scabs. Where trouble is caused by the blackening sticking to the tool, it is best to dust on a light coat of charcoal over the top of the heavy sticky blackening.   Charcoal dust is very light and is slow to absorb moisture; this makes it an excellent material to aid in the sleeking of sticky grades of blackening.   Where charcoal has been used, bellows are necessary to blow off all the dust that does not adhere to the surface of the mold; if this is not done, the loose dust will run before the metal when the mold is poured, and gather in lumps.   In fact, where it is desirable to have clean, sound castings, it does no harm, whatever the grade of blackening may be, to use the bellows to blow off the dust, provided the face of the mold is not broken by the force of the blast.

It is chiefly in medium-weight and heavy castings that it is found necessary to sleek dry blackenings. In light work, another plan, called **printing,** is largely followed. This consists in shaking the blackening from a bag evenly over the whole surface of the mold and then setting the pattern back carefully into the mold; the pattern is then rapped down lightly over the whole of its surface and in this way pressed into the blackening dust. It is again rapped lightly (to loosen it in the mold) ánd then withdrawn. If the above is properly done, the loose blackening dust will have been pressed down solidly on the face of the mold and will give form to the most delicate imprint of the pattern. In printing patterns, the molder generally has at least two bags: one holding a heavy blackening that he will shake on first, the thickness of this first coat being sufficient merely to cover the face of the mold about $\frac{1}{64}$ inch only; the other bag containing charcoal dust, or some other light grade of a specially prepared blackening. As soon as the dust from the first bag has settled, the second bag will be used, after which the pattern is *printed back*, as described. In doing this, the pattern must be perfectly dry. This is very important, for if there is the least moisture about the pattern, the blackening will stick to it when it is withdrawn from the mold. After a pattern has been imprinted, bellows are often used to blow off any blackening dust that may not have been firmly pressed on to the surface of the mold.

SKIN-DRIED MOLDS.

28. General Remarks.—Many large castings that it was formerly thought impossible to make except in dry-sand or loam molds are now made in green-sand molds by **skin drying.** Skin drying is also practiced with lighter work for the purpose of giving green-sand castings the surface and color of dry-sand ones.

It may be advisable to skin-dry some molds because of the nature of the sand used; the sand may contain too much clay, or it may be of such a character that it would not otherwise withstand the heat and wash of the metal. The purpose of skin drying is to give green-sand molds a hard surface, devoid of moisture as far as possible and similar to the hard and dry surface found in dry-sand and loam molds. For this purpose special physical characteristics are required in the sand that is used for the *facing*, as common heap sand can be used only for the *backing*. The facing sand should be of a loamy open nature, hardening only when heated, and also sufficiently porous to permit the metal to lie against its surface without bubbling or boiling. When unable to obtain the right grade of sand for making facing, the ordinary grades may be used if mixed with flour, molasses water, or clay wash. When flour is used, the usual proportions are 1 part of flour to from 20 to 30 parts of sand, according to the nature of the latter. When flour is used in the sand, care must be taken in drying the mold, for if the heat is great enough to burn the flour, it will cause the surface of the mold to crumble. The molasses water or clay wash may in some cases be used for wetting sand that has been mixed with flour, or the flour may be omitted and the sand sufficiently strengthened by the aid of the washes. Again, some sands, on account of their closeness, should be mixed with a sharp sand. Some localities possess molding sand naturally adapted to skin drying, while others do not, and, therefore, in the latter, more or less *doctoring* will be necessary to make the sand serviceable. The thickness of the facing used against the pattern generally ranges from 1 to 2 inches. After the facing sand has been banked against the pattern, common heap sand is used for a backing, and the mold rammed in the manner generally followed in green-sand work.

The facing for skin-dried molds is, as a general rule, used a little damper than facings would be for common green-sand work.

In cases where the sides of molds are very deep, and very heavy or deep copes are used, it is often well to ram up ½-inch to ¾-inch rods with every other course, as at Nos. *1* to *10*, Fig. 36. In this illustration the pouring basins are shown

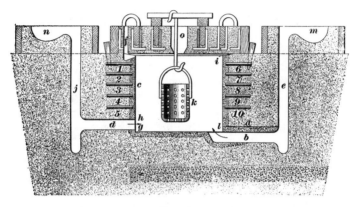

FIG. 36.

at *m* and *n*, the upright pouring gates at *e* and *j*, the inlet gates at *b* and *d*, the riser at *f*, the feeder at *o*, the fire-kettle at *k*, the dried crust at *c*, the lower corner at *g*, the inlets at *h* and *l*, and the inlet-gate top core at *a*.

Where the copes are skin-dried, they should, as a general thing, be very closely gaggered; and with some grades of sand, the surface should have nails between all the gaggers, with the heads coming even with the face of the mold and covered only with the blackening; for if this is not done, the dried crust on the surface of the mold may drop off easily. Not only is this practiced with copes, but in some cases molders will nail the side of drags that are over 6 inches deep as a protection against the dried crust falling away from its green backing. The gates and portions of the mold where the metal first enters are generally the parts that should be well nailed, for in skin-dried molds, if the surface once becomes broken, it soon washes the crust away, after which the material offers no greater resistance to the rushing metal than would so much dry dust.

In finishing the joints of skin-dried molds, it is essential that they be shaved as shown at *i*, Fig. 36, so as to prevent their crushing when the copes are closed, for the least pressure on the joint at the edge of such a mold may readily cause a **crush.**   There is no class of molds that requires more delicacy in handling, for the surface is only a crust about ½ inch thick that has but little union with the body of the mold, and may easily be separated from it by any jarring. Some molders will not trust to the nails for holding the portions of the mold surrounding the gates, but instead make cores the shape of that part of the mold and ram them up with the pattern.   This is the best method for preventing skin-dried molds from cutting at the gates.

29.   Finishing a Skin-Dried Mold.—After the mold has been made and its surface nailed over, as just described, it is finished by wetting the entire surface with molasses water.   This is done lightly by means of a camel's-hair, or other soft, brush.   After the surface has been moistened, it is sleeked up with finishing tools, as in finishing any green-sand mold.   Where molds are very large, they are moistened and finished in sections, for if all the surface of a large mold were moistened at one time, portions of it would be too dry to finish well by the time the molder reached them; when there is more than one molder working on a job, this precaution may not be necessary.   The mold having been sleeked with the finishing tools, the next proceeding is to blacken its surface, which may be done in one of two ways. One is to blacken the mold in the same manner that a dry-sand or loam mold is blackened; the other is to rub the blackening on dry in the dust form and then, after sleeking it as described in Art. **27,** to moisten the surface heavily with molasses water, applying the liquid with a soft brush. Rubbing the blackening dust on is only necessary on the sides of the mold, as a bag can be used to shake it on the bottom.   These two methods of blackening may often be used to advantage on the same mold.   The plan of rubbing the blackening on dry and then going over it with the molasses water, does not dampen the surface as much as when it is blackened with an all-wet blackening, as in the case of a dry-sand or loam mold. The reason why these two plans of blackening will sometimes work well together is because in skin drying the mold with fire-pans or sheet plates, there are some parts that will receive more heat than others.

By exercising care and judgment in dampening the sand and in blackening, all parts of the mold may become dry at about the same time; if the work is done in such a way that one part dries before the others, it may burn. In blackening the surface, it should be done as smoothly as possible, so as to avoid the necessity of much sleeking. By sleeking the wet blackening, a smoother casting may be produced; but unless it is carefully and skilfully done, there is more or less danger of the sleeking causing scabs. In putting the blackening on, it can be used thin enough not to be streaked, and with care in using a camel's-hair brush, no streaks need be shown, so that the castings can be made nearly as smooth as if the blackening were sleeked, and the danger caused by sleeking can be avoided.

30. Drying a Skin-Dried Mold.—In drying these molds, considerable judgment is required, for a scheme that will work well with one mold may not answer for another. That method must be adopted which is best suited for the work in hand. For example, some molds, such as those for anvil blocks, etc., may be dried by setting a fire-pot in them as shown at k, Fig. 36, where the cope is shown being dried above the drag. Sometimes the mold may be of such a form as to require a flat or square pan instead of the cylindrical kettle here shown; and with some molds this plan will not answer at all, because the mold is so shaped that kettles or pans cannot be used in them. These molds may be of such a form that their surfaces can be dried by laying sheet-iron plates, perforated with small holes, over them, and placing a fire on the plates; but this is a plan rarely used where kettles or pans can be employed.

The fuel commonly used in these appliances is charcoal; the fire should be mild and steady, especially at the start, since too strong a fire is apt to blister the face of the mold. Sometimes the cope and the drag may be dried together by having the cope propped up clear of the drag, and then heating between them by means of fire on perforated plates or in pans. Again, the mold may be such as to permit it to be closed while being dried, the riser and gates being left open to let out the steam, as shown in Fig. 36.

Natural or artificial gas may be used for drying molds. The gas is conveyed to the mold in a rubber tube having a piece of gas pipe in the end, and then burned against the face of the mold.

Green-sand cores or bodies forming the interior of molds are generally skin-dried by placing them in an oven and keeping the heat mild and uniform. To ascertain if a mold or core is skin-dried deep enough, it may be tested by cutting a small hole in the surface or by pressing the surface with the fingers. The most difficult places to dry by means of kettles or pans are the corners of a mold, as shown at g, Fig. 36. The sides of some molds might be baked and the binding material burned to ashes before the corners are dried. To get the corners dry, it is often necessary, after a kettle or pan fire has been taken out, to place hot coals or hot irons around in them to get them dry. It is here that the advantage of gas or hot air is apparent, as by either of them the heat can be directed to any given spot until it is thoroughly dried.

Any one wishing to acquire skill in skin-drying molds should begin on a small scale, as he is liable to make many mistakes at the start.

31. Gas Burner for Drying a Mold.—A good plan to follow when drying molds by means of gas is to substitute a Bunsen burner for the pipe. The objection to using an open pipe burner in the foundry is that usually a ⅛-inch or ¼-inch pipe is used, which is not only wasteful of gas, but its flame deposits such a heavy coating of soot on the face of the mold that the drying is not accomplished as fast or as economically as it should be with the amount of gas used.

The burner illustrated in Fig. 37 is a Bunsen burner that has been designed and used for this class of work; it is also well suited for any work requiring a gas heater. As ordinarily constructed, the burner consists of a piece of $\frac{1}{2}$-inch pipe a, 6 inches long, threaded at one end and screwed into a reducer b. The reducer has five $\frac{1}{4}$-inch holes c drilled into it to admit air. More holes may be drilled if required; and if, on the contrary, it is found that too much air is being furnished, some of them may be plugged up with wood, or, better, by tapping the extra ones and putting in $\frac{5}{16}$-inch screws. A $\frac{1}{4}$-inch pipe d, about 2 feet long, is threaded on one end and closed with a plug e driven or screwed into it. A hole is drilled lengthwise through the plug e, and a tube f of $\frac{1}{16}$-inch bore fastened into it. This $\frac{1}{16}$-inch hole fur-

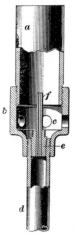

FIG. 37.

nishes the right proportion of gas to the burner. The tube f should extend at least $\frac{1}{4}$ inch past the holes c through which air enters. The lower end of the pipe d is connected to the gas-supply pipe by means of a rubber hose. When constructed in this manner, the burner may be used upside down or in any other position. The air and gas mix in the pipe a and the gas is completely consumed, because there is plenty of air mixed with it before it begins to burn. It gives no light, the flame being blue, but a great deal of heat.

32. Gating a Skin-Dried Mold.—In gating skin-dried molds, the method that will cause the least friction between the flowing metal and the surface of the mold is, as a rule, the best one to adopt. In Fig. 36 are shown two methods of gating that can be used with a large variety of molds. With a gate like that marked d, the metal will flow in as shown by the arrow h. Such a gate as this will cause great friction between the metal and the face of the mold, and unless the whole surface fronting the inlet gates is nailed very closely, with the heads of the nails even with the bottom face of the mold, the casting will scab at that point. Instead of nails, strong cores may be used to form all the surface fronting the inlet gate, thus preventing scabbing at that portion of the mold.

The best kind of gate to use for such work is shown at *b*, on the right of the illustration. Here the metal, on entering the mold, will come up from the bottom, as shown at *l*, and flow gently over all the face of the mold, causing little or no friction that might cause scabs. This form of gate is easier on the face of a mold than any other. It can be applied to a large class of molds. The only objection to it is that it does not distribute the dirt created in the pouring runner and gates or that may come from the scum of the ladle. As a rule, all such dirt will collect in a body and float right above the inlet *l*. In molds having cores or projections that will catch dirt and confine it in the parts especially requiring solid metal, this class of gate *b* would be undesirable. With an inlet gate as shown at *d*, the dirt is divided into fine particles and distributed to all portions of the casting, which, in some cases, may be preferable, even though there is a scab created in front of the gate.

GATES FOR MOLDS.

33. Pouring Gates for Catching Dirt.—The gate *b*, shown in Fig. 36, is very apt to collect and hold the dirt in one spot, but there are methods in use that serve to lessen considerably the amount of scum or dirt passing into these gates. Fig. 38 shows some of these methods, which may be modified as deemed desirable. In pouring a casting with a system of dirt catchers (commonly called **skimming gates**), shown in Fig. 38 (*a*), the metal first flows into the depression at *d*, filling it so that the core *h* holds back much of the dirt or scum coming from the ladle. Comparatively clean metal should pass from *d* down *e* through *g* to *f*. In the plan (*b*) it will be observed that the connection *g* between *e* and f is led to one side of the latter, so that the metal is given a whirling motion on entering f, which causes the scum and dirt to rise up into f, and causes comparatively clean metal to pass through e' into the mold; e' here corresponds to the inlet gates b and d in Fig. 36. To make this form effective, the inlet gate e', Fig. 38, which leads the metal into the mold, must have a smaller area than either

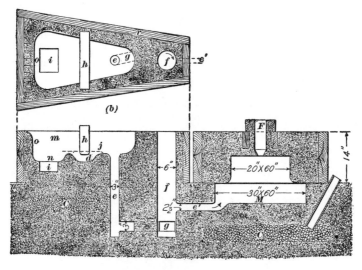

(b)

(a)

Fig. 38.

of the other openings. This is necessary so that the flow through e' will be *dammed* back and thus keep the riser f full of iron. Keeping f full of iron causes the dirt to float on top of it; whereas, did the metal in it descend to the level of e', this dirt would then pass into the mold. The drawing in Fig. 38 is not to scale, and the gates, pouring basins, risers, etc. are magnified in order to illustrate their relation to the mold.

The sizes of the various gates are given so as to give an idea of proportions that work well. The parts of Fig. 38 that are lettered, but have not yet been referred to, are m, the pouring basin; i, a core; r, the runner box; and F, the feeding head. With such a system of skimming gates, the under inlet gate b, Fig. 36, can be used with very little risk of having much scum or dirt pass into the mold. If the molder desires to decrease his labor in making such a system, the core h and depression d, Fig. 38, can be omitted and a level bottom used, as indicated by the dotted line j.

A study of Fig. 38 will show that intricate work is involved in this system; and that unless the molder exercises care and skill, there will probably be more dirt created by the sand washed from the corners and surface of the gates than would have flowed into the casting had there been but one straight gate and no skimming gates at all.

34. Skimming Gates for Medium and Light Castings.—It may be stated here that the arrangement of skimming gates is based upon the principle that all scum or dirt has less specific gravity than iron, for which reason it will float to the highest point that it can reach. In arranging skimming gates, some part is constructed to catch and hold the dirt as it rises to the surface of the flowing metal before it can enter the mold. The skimming gates just described

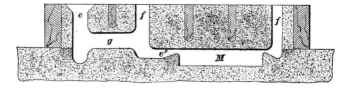

FIG. 39.

are for heavy castings; Fig. 39 illustrates a method suitable for medium and light castings. The sand is rammed around two sprue pins to form a pouring gate e and dirt riser f; a dirt-collecting channel g is then cut between them. The higher and longer the channel g can be made, the better dirt collector it will be. The metal having passed g flows on through e' to the mold M. In flowing from the gate g to the mold M, the scum or dirt in the metal, as it rises, is caught and held in the upper part of the channel *g* and dirt riser *f*. Sometimes the channel *g* is cut on a straight line to *f*, and sometimes on a curve, as at *g* and *f*, Fig. 38. If this is done properly, lumps of dirt should be seen whirling on the top of the metal in the dirt riser *f*, Figs. 38 and 39, and when breaking the channel *g* after the metal has cooled, dirt should be found in its upper part.

It is often desirable to have skimming gates for light work arranged so as to save as much labor as possible. This may be done by using the appliance shown in Fig. 40 (*a*), which is

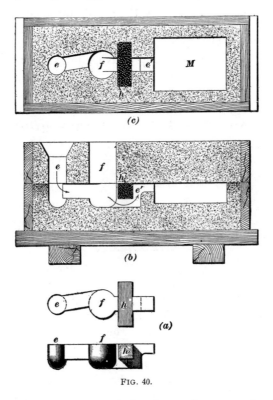

FIG. 40.

a small pattern so arranged that it can be rammed with the pattern in the nowel of small boxes or snap flasks. Two patterns or sprue pins are used to form the sprue *e* and riser f. When the mold is finished, the skimming gate should appear as shown in views (b) and (c), a small core being used as shown at h. The metal on being poured into the gate e flows into f with a whirling motion, and in going to the mold passes under the core placed at h, whence it passes through the gate at the entrance e' into the mold, as shown by the arrow.

The advantage of this skimming gate is that it can be formed very easily by means of the core h, which is set into the mold after the skimming-gate pattern is drawn. This core, being arranged to settle deeply into the gate, causes the iron that enters the mold to be taken from the lowest point of the skimming gate, which insures clean metal going into the mold. It is a simple device, but very effective in its results.

35. Top-Pouring Skimming Gates.—Another form of skimming gate is that used in **top pouring,** shown in

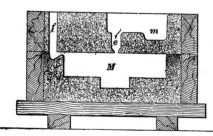

FIG. 41.

Fig. 41. This is applicable to a great variety of work in both light and heavy castings. In con- structing such gates, a pouring basin m is made; the metal en- ters the mold through a gate or gates e. By a quick dash of the metal from the ladle at the start when pouring, and by then keeping the pouring basin full until the flow-off risers show that the mold has been filled, the dirt will stay on top of the metal in the pouring basin, leav- ing clean iron to pass into the mold. It is only when begin- ning to pour that dirt should have any chance to enter the mold; and this is true to a greater or less extent of all forms of skimming gates. There are many forms of skimming gates, but those shown in Figs. 38 to 41 should be sufficient to indicate the principles involved.

Probably the easiest way to obtain clean castings is to keep the pouring basin full during the time of pouring. In these illustrations the gates are shown large in proportion to the balance of the mold in order to show their arrangement more clearly.

36. Dirt in Castings.—After scum or dirt has entered a mold, it must locate itself somewhere. Its natural tendency is to rise to the top of the mold, since it has a lower specific gravity than iron; but there are conditions that at times prevent this. When iron enters a mold, it rapidly loses its fluidity, and for that reason if dirt drifts to the side of the mold, it is liable, on account of the

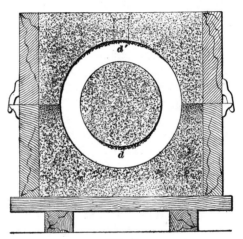

FIG. 42.

dulness of the metal, to stick there and let the metal flow over it. Again, molds often have projecting cores that, when the metal rises to the under side, catch the dirt and retain it.

If we consider the case of cylinders or pipes cast horizontally, we shall find that the scum or dirt will lodge as seen at d, Fig. 42, and what passes this point will rise to the top and stay at d'. On this account it is often necessary, if the inside is to be bored out, to leave extra stock for finishing at d. If the castings are columns for supporting buildings, or similar pieces, it would be wise to make them thickest on the top or cope side at d', to allow for the weakness that the dirt might cause at that point. Even with the same amount of dirt at d and d', and with the nowel and cope parts of the casting of the same thickness, the cope will be the weaker side of the casting, as there is less pressure at that point of the mold to make the metal solid.

37. Pouring Basins.—In making a mold, there are few things that require greater care and skill than the **pouring basin.** Here the greatest amount of friction, rush, and washing-out effect of the metal is produced; and if the basin is not well made, it will be easily cut by the falling metal. If the basin once starts to cut, considerable damage may result before the mold is filled. A molder may slight the rest of his mold and yet have his castings come out so that they will pass inspection, but any carelessness or ignorance in making basins, runners, or gates will cause trouble. In pouring a mold, the iron first drops from the ladle into the basin, from which it runs with more or less velocity into the upright sprues or runners, and from them into the gates that lead into the mold. With the exception of that portion of the mold into which the iron enters or drops, there is very little agitation of the metal as it gradually rises in a mold, compared to the rush and spattering that exists in the pouring basin. When making these basins, extra care should be taken to see that the sand has been well mixed and riddled before it is shoveled into the basin box. The use of poorly tempered sand for making basins has often caused bad castings. Some molders shovel a little sand into the box to form the bottom of the basin and then tramp it with their feet or pack it with a rammer, after which they press sand against the sides of the box with their hands to give shape to the pouring basin; this is a very bad practice to follow. To make a reliable basin, the box should first be evenly rammed full of sand, after which the shape of the basin can be dug out with a shovel or trowel. The ramming will give a firm, solid body to the sand. The point of danger in pouring basins is at the bottom n, Fig. 38. Here the force of the dropping metal has such a cutting effect that in the case of large basins, it is advisable to place a core at i; in some cases bricks may be used instead; or again, green sand may be made to take the place of the core or bricks, by closely nailing the bottom where the iron will drop from the ladle, the heads of the nails being left even with the surface of the sand. It will be noticed that a well is formed, around the core i, so that immediately after starting to pour, a body of metal will be formed into which the iron drops, and thus save the bottom of the basin n from receiving the full force of the dropping metal. In some cases the well at n can be made so deep that there will be no necessity for cores or bricks. However, it is well to secure this part of a basin as much as possible, for a sudden jerk of the ladle at the start, which often occurs, might prove very serious and result in the loss of the casting. There should never be less than 4 inches of good tempered sand between the bottom of the basin at n and the floor, or other bottom. In the case of very large basins, it is wise to have a cinder bed under them, as shown at C, Fig. 38.

If wooden basin boxes are used, it is well to have their fronts nailed, as shown at o, Fig. 38 (b), as the wash of the metal striking the front of the basin has been known to cut away the sand and cause a bad casting. Another point to be carefully watched in making pouring basins is to avoid having water carelessly swabbed around the edges of the gates or the bottom portion of the basin, as this may start the metal blowing, and when this has once commenced, it is hard to tell when it will stop. Any boiling of the metal in the basin will create more or less scum or dirt that must follow the metal through the gates into the casting.

GREEN-SAND MOLDING.

(PART 4.)

IRON AND BRASS MOLDING.

CHAPLETS.

1. Types of Chaplets in General Use.—Chaplets
are used to support or hold down cores that, owing to their
shape, are not self-supporting when placed in the mold.
There is a large variety of such chaplets in use; those shown
in Fig. 1 are called **single-headed chaplets,** and those in
Fig. 2 **double-headed chaplets.** Fig. 3 (*a*) is a *spring
chaplet;* (*b*) is a combination of *chaplet and stand,* a scheme
often used to save labor and time in setting chaplets. The
stand can be set in the nowel under the pattern when it is
being rammed up, and then when finishing the mold, the
chaplet is set in place. Often iron cross-bars in both drag
and cope are cast with bosses into which holes are drilled to
allow chaplets to be inserted after the manner shown in
Fig. 3 (*b*). The chaplet (*a*) in the same illustration is made
of a piece of steel or iron that can be sprung to the desired
form. There are places where these chaplets are of special
value.

In Fig. 1, chaplet (*a*) is made of round iron cut to any
desired length and having a solid head; (*b*) is a *chaplet stem,*
on which, at the head *o*, any size of plate may be riveted.

§ 43

Often larger heads are required than can be forged upon a chaplet, as in (*a*), and such chaplets as (*b*) can be fitted with a head of the size desired; (*c*) shows a stem with such a head *q* riveted on. There are times when the pressure on a chaplet will be so great, or when the cores require such a large bearing surface that the plate *q* on chaplet (*c*) would be

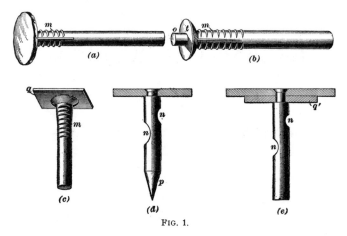

Fig. 1.

better if reenforced by a back plate, as shown at *q'*, chaplet (*e*). Chaplet (*d*) shows a sharp point *p*, which.is sometimes necessary when it is desired to drive the chaplet into bottom boards, wooden blocks, etc., underneath the surface of the mold. In driving such chaplets, a depth of $\frac{1}{2}$ inch to $\frac{3}{4}$ inch is sufficient, as the driving is liable to force down the block or to jar the board and loosen parts of the mold.

Referring to Fig. 2, (*a*) is a *double-headed chaplet* stem, of any desired length, provided with pins, to which plates of any size may be riveted, as shown at (*b*). Double-headed chaplets often have to be fastened to the surface of the mold or core, so that any jarring of either may not move them; this is done by making the chaplet with a sharp stem, as shown at *p*, Fig. 2 (*c*). These sharp stems are driven into the face of the mold; they seldom need to be more than $\frac{3}{4}$ inch long. In chaplet (*c*) the heads *q* and *q'* are placed on after the stem has been made and the top head *q* is riveted on.

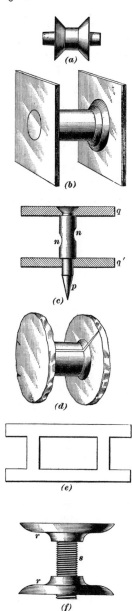

(a)

(b)

(c)

(d)

(e)

(f)

FIG. 2.

A double-headed forged chaplet is shown at (*d*), while (*e*) is a cast-iron one. Cast-iron chaplets can often be used, but they must not be placed where they will be struck by the stream of metal from the pouring gates, as they melt more readily than wrought iron. This is a point that must be observed in the use of all chaplets, as many castings have been lost because molders have thoughtlessly set chaplets in front of gates that deliver large bodies of metal; or, again, the quantity of the iron may be small, but so hot as to melt the chaplets.

There are several firms in this country manufacturing chaplets, especially those shown at (*a*), (*b*), (*c*), Fig. 1, and (*a*), (*b*), and (*d*), Fig. 2. Chaplets can be purchased so cheaply that any person requiring only a few cannot afford to make them.

Fig. 2 (*f*) shows an *adjustable chaplet*, a very convenient appliance where odd lengths are needed. It consists of a stud or stem *s*, made by threading stock of the required size in the screw machine and cutting it to the most convenient lengths. Ordinary cast-iron washers *r* are drilled and tapped to suit the threaded stem *s*. An adjustment $\frac{1}{2}$ inch, more or less, may be made with the washers *r*, while a variety of lengths of stems permits the making up of any size. The stem here shown is $\frac{5}{8}$ inch in diameter and the washers $2\frac{1}{2}$ inches.

2. Precautions in Using Chaplets.—Chaplets are generally "necessary evils," since they are very apt to weaken the casting more or less at the point where they are

(a)

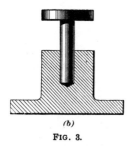

(b)

FIG. 3.

placed. This may be done in three ways : *First*, by breaking the uniformity of the metal of the casting by the introduction of some other substance at the point occupied by the chaplet; *second*, by working loose and leaving holes in the castings, generally caused by blowholes around them; *third*, by causing porous or unsound metal to form around them. The first of these evils cannot be avoided; but by good design and care in the making and use of the chaplets, the second and third evils can be greatly decreased, and in many cases almost wholly avoided. In regard to the second evil, chaplets should be nicked or have depressions made in them, as at n, Fig. 1 (d) and (e), or else have burrs on them, as seen at m, Fig. 1 (a), (b), and (c).

Some molders, in making the stem, avoid the heavy shoulder t shown in Fig. 1 (b) and make the stem sufficiently large to give a good bearing to the head q in (c), as shown in Fig. 1 (d) and (e). Some persons cut a thread on the part of the stem that is cast in the metal, even in the case of chaplets that have fixed heads. This scheme is used also in making double-headers, as in Fig. 2 (b), and in such cases a thread will be cut the whole length of the stem and the heads screwed on, as in Fig. 2 (f). The screw stem has an advantage in another way, as the cutting of the thread removes all scale or rust from the surface of the stem, and this is very important.

3. Rust on Chaplets.—There is always more or less **rust** or **scale** adhering to the surface of both old and new chaplets. As much of this as possible should be removed from the parts that are cast in the metal. When molten metal comes in contact with rust, a gas is created. It is calculated that 60 grains of dry rust will make 31 grains of carbonic-oxide gas, which at 2,800° F. (the temperature of molten iron) and the pressure of 1 atmosphere will occupy about 600 cubic inches of space. It does not require a very large piece of iron to give 60 grains of rust. The space that the gas occupies depends on the pressure; and the harm it can do depends on the rapidity with which the metal solidifies and prevents the gas from escaping. Blowholes are rarely found around the chaplets in the lower part of a casting; they occur in the upper part where there is very little pressure during the pouring. The part of a mold where the greatest pressure exists is usually the first to be filled, and the iron is also hotter and cleaner there than at the top of the mold. If, for any reason, the chaplets at the bottom should cause the iron to boil or blow, the gas will generally escape upwards through the metal and out at the top of the cope sand or out of the flow-off gates.

Chaplets may be free from rust when placed in the mold, but if kept there for two or three days before the mold is cast, they are very apt to become rusty, especially if the mold is a green-sand one. There are varnishes that can be used to prevent their rusting, and these will be dealt with further on.

4. Moisture on Chaplets.—A piece of polished iron, if exposed to moist air or otherwise moistened, soon becomes rusty. This is due to the affinity that iron has for oxygen. Under certain conditions, polished iron can be kept free from rust by keeping it in a dry atmosphere. Take polished iron from a cold room into a warm, moist one, and it will not be long before rust will be formed on it. This is caused by the cold iron condensing whatever moisture there may be in the air immediately surrounding it on its surface. The more rust there is on iron, the more moisture it will collect; and on chaplets, this moisture can do greater injury

than rust. To test this, take a rusty rod and heat it sufficiently to dry all its moisture, after which place it quickly
into a ladle of molten iron. The metal will bubble around
the rod, more or less, but it will not fly out of the ladle as it
would if there were moisture on the rod. The steam from
the moisture on chaplets may cause a great amount of
bubbling or blowing, as can be seen by quickly immersing a
damp rod in a ladle of iron.

The best thing that can be done to prevent chaplets from
becoming rusty and collecting moisture is to tin that portion
liable to be incased by the metal. Coating the iron body
with tin not only prevents the oxygen or moisture from
attacking the iron, but it has an affinity for iron that makes
the iron more fluid when in a molten state, and this greatly
aids the release of any gases that might be created around
the chaplets.

Where chaplets are not tinned, their exposed parts may
be covered with a coating of red lead mixed with turpen-
tine. Asphaltum, coal tar, and chalk are often used as a
coating. Where there is much moisture, these materials
may collect sufficient dampness to cause injury, but not to
such an extent as rusty chaplets.

5. Setting and Wedging Chaplets.—There are few
things more annoying than to see castings lost by thought-
lessness or ignorance in setting and wedging chaplets.
Chaplets are generally set into a cope by first passing a
$\frac{1}{8}$-inch rod or vent wire up or down through the cope at the
spot where it is desired to place the chaplet. If the chaplet
is larger than $\frac{1}{4}$-inch, then a $\frac{1}{4}$-inch vent wire or rod is passed
through the hole made by the $\frac{1}{8}$-inch rod, and so on, increasing the size of the rods according to the size of the chaplet
stem, the idea being to gradually enlarge the hole for admis-
sion of the stem without applying much pressure. After
the stem has been pressed through the body of the cope, it
should then be pulled out and the hole reamed out at the
face of the mold, as seen at c, Fig. 4. Many molders having
failed to do this when pressing down the chaplet to a good

bearing on the core, or in wedging it down to place, have caused the face of the mold around the bottom of the stem to be pulled down, as at *d*, thus resulting in the loss of the casting.

After the chaplets have been set and the cope closed, some molders drive in wedges with a hammer to fasten the chaplets, as at *e*. This results in displacing the chaplet, as shown. The chaplet should be placed solidly upon the core *A* and

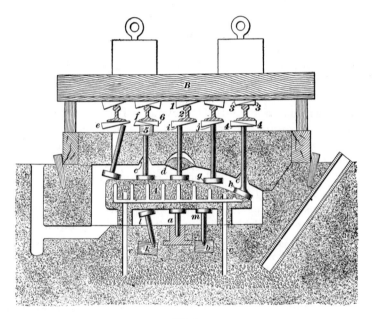

FIG. 4.

then wedged down, as shown at *f*. Then, again, some molders in fastening down cores having slanting surfaces use chaplets having the head at right angles with the stem, as at *g*. Where the surface of the core is slanting, the heads of the chaplets should be set on the stem at the same angle as that of the core; care should be taken that the slant of the head and core agrees, so that they may come solidly together when the cope is closed, as at *h*.

6. Blocking on Top of Chaplets.—Very often the chaplet stems will require some blocking on top of them, as at *5*, Fig. 4, before wedges can be used. Where this is necessary, one must be sure that the blocking is over the center of the stem and also that there is sufficient space to use the wedges *6*. While it is often necessary to use blocking, there are many cases where its use can be avoided if judgment is used in placing the rails, in weighting down the binder *B*, and in getting the chaplet stems the right length. Molders often leave the chaplet stems sticking up from 3 to 6 inches above the top of the cope; and sometimes they leave them from 1 to 3 inches below this same level.

Lack of system and judgment causes some molders to use from 100 to 200 pounds of small pieces of blocking to fasten down a half dozen chaplets, where others will do the same work without a pound of blocking. The fewer pieces of blocking that are placed between the weighting-down binders or rails and the wedges necessary to fasten the chaplets, the better it is for the safety of the mold. In many cases, with forethought and judgment the chaplets can be cut to such a length and the binders can be so arranged as to avoid the necessity of using any blocking between the top of the chaplet stems and the bottom of the weighting-down binders. Where this is possible it should be done; and in allowing space for wedges between the top of the chaplet stems and the bottom of binders, it should range from $\frac{1}{2}$ inch to $\frac{3}{4}$ inch. Before wedging the chaplets, the weights necessary to hold down the cope should be placed on it, as all the binders, etc. will spring more or less when weighted; and if the weights are placed after the chaplets are wedged, this springing may drive the chaplet heads into the face of the cores and cause damage. Just before pouring a mold having chaplets, it is good practice to go over it to test the wedges, as they sometimes work loose after they have been tightened.

7. Placing Chaplets in Bottom of Mold.—It is as important to have chaplets set correctly in the bottom as in the top part of molds. A large number of the chaplets used in the bottom of molds are driven into the bottom boards or into wooden blocks. Where wooden blocks are used to support cores of any great weight, it is best to make them of hard wood, and set them with the grain up. In driving the points, as at b and k, Fig. 4, the molder must use his judgment in driving them the proper distance, taking into consideration the size of the chaplet stem, the weight of the core, and the nature of the block.

Some molders lose castings by the manner in which they set bottom chaplets. The block k, Fig. 4, is apt to split for two reasons: *first*, because the chaplet has been driven in too far, and, *second*, because the point is near the end of the block. The block b shows the proper practice in these respects. Where points are driven as at k, castings are very liable to be lost through the settling of the chaplet into the block. Even if the weight of the core does not do it, the wedging down of the top chaplets when the cope is closed will probably do so. A fine wire should be pressed through the sand to find the end of each block, so that the chaplets may be driven near the center of each block. The head of chaplet l is shown in such a position that it makes an angle with the core and has but one edge touching it. It is as important to have bottom chaplets set properly as top ones, and it can readily be seen that a core should set solidly on the chaplet, as at m. At a is shown a chaplet placed in a stand, the appliance being illustrated in Fig. 3 (*b*). These stands are very good for some classes of castings, as, for example, in cases where it does not matter if the face of the casting is chilled a little at and around the spot of the chaplet connection. By placing sand in the bottom of the hole admitting the chaplet's stem, any variation in the thickness of the metal can be arranged for in setting the chaplet.

8. Wedges for Setting Chaplets.—In making wedges of either wood or iron, it is best to make them with as little taper as practicable. The greater the taper, the more difficult it is to fasten the wedges and the more liable they are to work loose. The wedges shown in Fig. 5 represent what is considered good practice. They can be fastened without much danger of their being loosened by jars when other wedges are being driven. Wedge (*b*) is generally used for

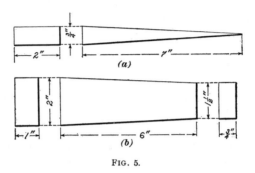

FIG. 5.

resisting great strains and to partly take the place of blocking; it can be made of any size to suit the conditions of the work. The quick-tapering wedge (*a*) is the one most commonly used, and the dimensions given can be applied to either cast-iron or wrought-iron wedges.

"DRAWING DOWN" OF THE COPE.

9. The phrase **drawing down of the cope** is applied to copes whose surface sand drops down upon the metal as the mold fills up. This is caused by the heat of the metal drying the surface of the cope. If the sand in the cope is not of such a nature as to bake solidly to the depth penetrated by the heat, and it does not hang well when one part is drier than another, then the sand will drop upon the metal in small quantities and cause lumps and dirt holes in the upper surface of the casting. This depends not only on the nature of the sand, but also on whether the mold is kept air-tight or not. With strong sand in the cope, some molds may be cast with their feeders, risers, etc. all open, and no injury will occur to the casting. If this were done with other grades of sand, the whole surface would be drawn down. It is chiefly with such castings as thick plates or heavy blocks, where the cope surface is exposed to the direct heat of the metal from the moment it enters the mold until it comes against the upper surface, that difficulty is experienced by drawing down. Any part of the cope's surface that is exposed for $\frac{1}{2}$ minute to the direct heat of rising metal should have a strong grade of surface sand for the first $1\frac{1}{2}$ inches. In addition it should be closely gaggered, with the sand not more than $\frac{3}{8}$ inch thick under the gaggers. The first course of sand in such copes should be evenly and firmly rammed; and the weaker the sand, the harder should be the ramming.

When the sand is weak, it may be strengthened either by mixing flour with it or by wetting it with clay wash or molasses water. After the surface of the mold is finished it should be sprinkled with molasses water; in fact, it is well to do this even with strong grades of sand, where they are expected to be exposed to the direct heat of the metal for more than $\frac{1}{2}$ minute. Use 1 part of flour to from 15 to 25 parts of sand, according to the strength of the latter; the weaker the sand, the more flour is necessary. In some cases where the copes are to be exposed to intense heat, as in very thick plates or anvil blocks, it is often advisable to nail all the surface between the gaggers, keeping the nail heads either even with the face of the cope and covered with blackening or else $\frac{1}{4}$ inch away from the surface and covered with sand and blackening. In venting green-sand copes that are liable to be drawn down, the vents should not be carried any closer than within 1 inch to $1\frac{1}{2}$ inches of the surface. For if they are carried close to the surface, they will permit the escape of gases and relieve the pressure against the face of the cope.

Drawing down occurs not only in green-sand copes, but also in dry-sand and loam copes or covers. Both weak and strong sands are used in these latter molds, as well as in the green-sand molds. Where there are heavy bodies of molten metal directly under the copes, dry-sand and loam mixtures should also be strong, or else drawing down will occur, as in green-sand work.

PRESSURE OF GASES IN MOLDS.

10. Object of Maintaining Gas Pressure in Molds.—There is a great deal of difference in the practice of molders in leaving the feeders and risers open or closed while casting. In most cases it is best to pour large molds with all the risers and feeders closed perfectly air-tight. Molds that are liable to have the copes *drawn down*, or in which the rushing of hot gases upwards through the risers will have a tendency to draw the gases from the vents in the bottom, should be cast air-tight as far as possible. The rush of hot air and gases through open risers has a tendency to divert the gases generated in the bottom of the mold from going downwards, causing them to pass up through the under surface of molds and producing scabs on the bottom of the casting. This rush also relieves the mold of the internal air pressure that exists when the mold is kept air-tight; and this pressure is often sufficient to prevent weak grades of sand from being drawn down from the surface of the cope.

Air is like all other gases in that it expands with an increase of temperature. At a temperature of 500° F., air has about double the volume that it has when at 0° F. The temperature of the air and gases in a mold is perhaps about one-half that of the rising metal; it may safely be taken as at least one-third. This means that the air and gases in a mold would have a temperature of 600° to 1,000° F., and that the gases in a mold having open risers would be increased in volume to two or three times that before the liquid metal commenced filling the mold. In an air-tight mold this would cause the pressure of the air to increase with its temperature. In other words, the air in a mold before casting was begun would have the pressure of the atmosphere, which is about 14¾ pounds per square inch, but by increasing the temperature, the pressure of the air and gases in an air-tight mold would increase from 2 to 4 pounds per square inch over the atmospheric pressure. Such a pressure at the face of the mold is very effective in preventing the surface of the cope from drawing down, and also prevents the gases from rising in the lower part of the mold where they might cause scabs, or worse still, start the mold blowing. In casting small molds the risers can sometimes be kept open without injury; this is also true in dry-sand and loam molds that have solid faces with little cope surface exposed to the metal. Aside from these, all risers and feeders should, as a rule (especially in heavy green-sand work), be closed air-tight and weighted down so that any increase of gaseous pressure cannot lift them.

Where risers are left open in any mold, they should be sufficiently large in area to allow the expanded air and gases to escape freely, as air rushing through these passages is apt to do damage by cutting away the sand around them.

11. Coverings for the Feeding Heads and Risers. When covering the feeders and risers, as at *n* and *o*, Fig. 6, care should be exercised, as the smallest opening leaves room for the air and gases to rush out. This would wear the sand

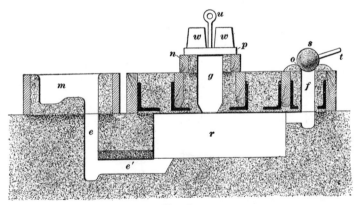

FIG. 6.

and increase the area of the opening for escape as long as the mold was being poured. In placing covers on the feeding heads, as at *n*, the covering plate *p* should have a good bearing over the top of the feeder box or projection, and to insure a tight joint, flour paste, parting sand, or dry flour is put around the opening for the covers to bear on. Flour paste is the best, and should be used for all heavy castings or work having cope covers or projections where the liberation of compressed air or gases might cause the molds to scab or blow. After the feeding head g is covered it should be weighted as shown at w. The combined weight of the cover and weights placed on feeders or risers should be estimated according to the possible temperature of the air and gas inside. This point must be left largely to the judgment and experience of the molder. It is always best to have more weight on p than is really necessary, for an excess of weight can do no harm. Where feeders or risers are not more than $2\frac{1}{2}$ inches in diameter, balls of clay are often used, as shown at s. These balls of clay should be of such consistency as to hang together well. In some cases it is well to stick a rod in the ball of clay, as shown at t, and to have handles on the cover-plates, as shown at u.

BLOWHOLES.

12. Causes of Blowholes.—In pouring a casting with very dull iron, it is sometimes advisable to leave some risers

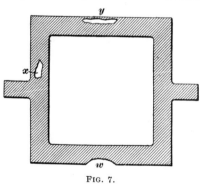

FIG. 7.

open. Where there is no great danger of the cope drawing down or the gases working upwards from the bottom of the mold this is desirable. Too close confinement of the air and gases may cause blowholes in the thinner parts of the casting. Castings frequently have what are called *shrink holes* and *blowholes*. The latter are generally holes having smooth surfaces and are rarely larger than would hold a teaspoonful of water. Castings may be so full of blowholes as to look like a honeycomb.

Blowholes, whether in large clusters or singly in a casting, are caused by the gas or steam (generated from the
moisture in the sand, facings, etc., that compose the mold)
endeavoring to escape. The gas or steam is caught because
the metal is dull and solidifies before the gas can pass out
entirely. Blowholes may be expected in any casting where
the cores do not vent freely, or where the mold, for any reason, may cause the metal to kick or blow. Where the metal
kicks or blows in a mold, the formation of blowholes may
often be prevented by flowing metal through the risers.
This is especially effective where the inlet gates carry the
metal to the bottom of the mold, and flow-off gates or risers
are placed at the top. Where hot metal fills a mold that
blows or kicks and is slow in solidifying, the gases may free
themselves without the flowing of much metal through the
risers.

There are two kinds of blowholes: one is found in the
interior of castings, as at x, Fig. 7; the other is found on or
near the exterior, as at w and y. These exterior holes are
more often found in the form of indentations, as shown at w.
Such a formation on the bottom of a mold or casting may
be caused by the heat of the metal drawing the gases to one
spot. Gases are liable to collect when there are several
vents leading to one main vent, or where there is a softness
in the mold at that point. Where there is a hard spot in
the face of the mold, the gases, not finding relief downwards,
try to pass through the metal and are caught and imprisoned
by the solidifying iron.

The form of blowhole seen at y, Fig. 7, is generally caused
by gases passing from the bottom of molds or cores upwards
through the metal in an effort to escape through the cope
surface. If there is a damp or hard spot in the surface of
the cope, it will chill the metal and form a thin crust through
which the gases cannot escape, and being unable to go any
farther, they will be imprisoned and form a hole, as shown.
Such holes as at y are chiefly found in thin castings, ranging
from $\frac{1}{2}$ to 1 inch in thickness; in such cases the cope exerts
a greater chilling effect than it does where there is a thick

body of metal. Often the surface of a light casting appears solid until one passes a scraper over it, when, from the sound made, hollows may be detected. On breaking the crust, it will be found that these hollows or indentations are generally of a very smooth character, which shows that they were formed by imprisoned gases. A remedy for this is to have vents in the bottom of the molds of such a character as to allow the gases to pass off freely in that direction, and also to guard against copes having wet or hard surfaces. The drier the sand can be used and the softer it is rammed at the surface of copes covering thin castings, the better.

SHRINK HOLES.

13. Causes of Shrink Holes. — A **shrink hole** differs in appearance from a blowhole in that the former

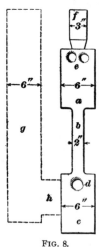

FIG. 8.

generally has a rough surface while the latter has a smooth one. The shrink hole generally looks as if a body of metal, of the same form as the hole, had been torn out, leaving a very rough open-grained fracture. It is caused by the parts that are the first to solidify shrinking and drawing metal from those that solidify last. This is due to the peculiarities in the cast iron that cause expansion at the moment of solidification, just before contraction takes place. It is also effected by the fact that thin or exterior portions solidify with a closer grain and possess more combined carbon and a greater specific gravity than interior or heavy parts, even when poured from the same ladle of iron. The harder the iron, the more noticeable will be these conditions. With very soft grades of iron little difficulty will be experienced from shrink holes, unless the castings are heavy. Where the iron is hard, either high in combined carbon or low in graphitic carbon, the shrinkage will always be great compared to that in soft iron. This is so true that, unless hard-iron castings are very carefully proportioned, considerable shrinkage may be expected in the parts that are the last to solidify. These should be provided with feeders through which compensation for the shrinkage is made with good hot metal. This point is well shown in Figs. 8 to 11.

Molders are sometimes held responsible for shrink holes that it was impossible for them to avoid. For example, consider Fig. 8, which shows a section of a casting having a light body at *b* connecting two heavy ones, as at *a* and *c*. It is evident that the lighter body *b* will solidify before the heavier ones *a* and *c*. This means that such a piece, if cast in a vertical position, will probably have shrink holes at *d*. The reason for this is that after the light section *b* has solidified, the outer portion of *c* will draw metal from the portion that was the last to solidify, forming cavities, as shown at *d*. The holes *e* occur near the upper end of the casting, instead of near the light section *b*, owing to the fact that the metal commences to solidify at the bottom and then gradually works upwards until the whole body is solid. The heat in escaping from a casting moves upwards more freely than in other directions, and hence the topmost bodies are kept hot the longest. The holes at *e* in the upper end of such a casting can be prevented by having a feeder *f* through which, after the casting is poured, additional metal may be fed to replace that taken away from the portion at *e*. The only way to prevent holes at *d* in such a vertical casting is to have a feeder leading down to *c*, as indicated by the dotted part *g*, when this is practicable; but this is seldom possible in ordinary work. It should also be borne in mind that if the feeder *g* is to be effective in preventing holes *d*, this feeder must be larger than the section *c*, so that the feeder and its inlet *h* will be the last to solidify.

Another example of a shrink hole may be seen at the upper end of the vertical casting shown in Fig. 9. Here the riser head *f*, used chiefly to receive the dirt, has an area at *a* larger than at any other portion of the section. A study of Fig. 8 should make it clear why shrink holes are formed at *a*, Fig. 9. To prevent the formation of such holes, perfect feeding is required. If the feeding is omitted, then the section at *f* should be enlarged to a thickness equal to that of the casting measured on the line *m m*. This would

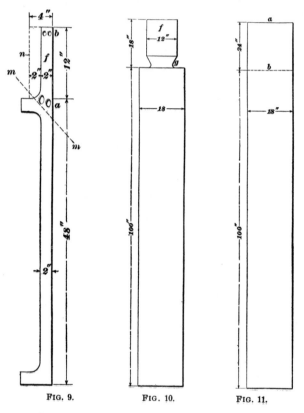

FIG. 9.   FIG. 10.   FIG. 11.

give us a *sinking head*, like that shown by the dotted line *n*, 4 inches thick instead of the original 2 inches. By having a sinking head 4 inches thick, it will be self-feeding, so that instead of the holes being at *a*, we shall find them higher up, at *b*. The principle involved in this scheme of self-feeding is also illustrated in Figs. 10 and 11. Here is shown a roll supposed to be cast on end. Some molders, to save lathe work or labor in cutting off feeding heads from such castings, will place feeders as seen at *f*, coming down to a neck, as at *g*, and then feed the casting with churning rods and an occasional pouring of hot iron into the feeding head.

Instead of depending on care, judgment, and skill in feeding such castings, some molders avoid the neck at *g* and simply carry the casting up straight, and have it appear as shown in Fig. 11. After these molds are cast, the upper end is covered with blackening to prevent heat escaping, and then the extra length of the casting used for a sinking head is kept full of metal, by occasional pouring as it shrinks away. After the casting is cold, the extra length provided for a sinking head is cut off. The length of the sinking head from *a* to *b*, in some cases, ranges from 2 to 3 feet, in order to insure the body of the casting below the level of *b* being perfectly solid. A great many founders cast hydraulic cylinders, rolls, shafts, cannon, etc., on this plan, having learned from experience that this is the best way to obtain castings free from shrinkage defects.

SHRINKAGE AND CONTRACTION.

14. The actions of *shrinkage* and of *contraction* are distinct in their nature, and are separated by the act of expansion that takes place at the moment of solidification. **Shrinkage,** as here considered, is that property of the metal when liquid that causes it to decrease in volume while cooling, until the moment it becomes solid. At the moment the iron becomes solid there is a slight expansion, but as it is much less than the previous shrinkage, it is rarely noticeable, especially in large castings. After iron has solidified, it again decreases in dimensions until it reaches the temperature of the atmosphere; this is **contraction.** While this distinction is not always recognized, yet some such distinction is necessary, and these terms will be used as defined. Light-work molders and founders not having to make heavy castings, in which the greatest shrinkage is displayed, are very apt to confound shrinkage with contraction. Nevertheless, there are two separate actions, and they should be recognized by distinctive terms.

FEEDING THE MOLD.

15. Importance of Proper Feeding.—Feeding can be properly done in many cases without the use of large **feeding heads.** One objection to large feeding heads is that it is expensive to remove them from the casting. Many molders think that as long as a casting has a feeder, no matter how small it is, that is all that is necessary. The man that thoroughly understands the principle of feeding will proceed with discretion in placing such feeders as shown in Figs. 10 and 11. If he has a feeder like that in Fig. 10, he knows that great care and skill must be used in order to get a sound casting. In contrast to this, it is no uncommon thing to see a molder using a feeding head that will solidify almost as soon as the mold is poured, although the casting underneath the feeder may remain in a fluid state from 15 to 30 minutes. Such castings would be better without any feeder, as by a small feeder a molder can draw out more iron than he puts in. Whenever a feeder is used, it should be sufficiently large to permit of feeding the body of metal below it as long as the latter remains in a fluid state. Some molders accomplish this by means of a smaller feeder than others, although the difference in size may not be great.

Usually, the greatest difficulty in feeding is to keep the smaller sizes of feeding heads open. A head from 2 to 3 inches in diameter should be kept open from 10 to 15 minutes, and the molder should be careful not to use too large a feeding rod. Sometimes a molder will take a $\frac{3}{4}$-inch round cold-iron rod and use it in a 2-inch feeder. The result is that the cold rod chills the liquid metal the moment it is inserted, and being too large for the feeder, the chilling rarely permits the removal of the rod.

16. Use of Feeding Rods.—For a 2-inch feeder the diameter of the **feeding rods** should not be more than ⅜ inch. For a 3-inch to 4-inch feeding head, a ½-inch rod will work well. For 4-inch to 6-inch feeders, ⅝-inch to ¾-inch rods will work nicely. In all feeding heads above 6 inches in diameter, rods from ¾ inch to 1 inch in diameter can be used. Before being inserted into a feeder, the rod should always be well heated in a ladle of hot metal, so as to prevent its chilling the metal in the feeding head. The metal in the feeding head is generally dull on account of its having

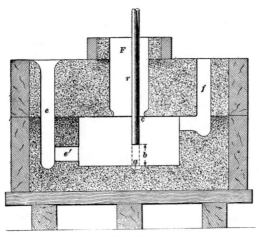

FIG. 12.

flowed upwards from the bottom to fill the feeder. When placing a rod in a feeding head, it should pass through the head and enter the casting to a depth that will insure hot iron being fed into the part that is going to shrink. Too many molders push the feeding rod no deeper than the bottom of the feeding head. Some do this through ignorance, and others to get rid of a hot job quickly. In the ordinary molds, one should, on inserting the feeding rod *r*, Fig. 12, be careful to pass it down gently until it is felt to strike the bottom of the mold, and then to draw it up a few inches, as shown at *b*, so as not to make a lump on the casting at *a*.

After the feeding rod has been raised off the bottom, it should be worked up and down, or churned, through a distance of from 3 to 5 inches, according to the character of the mold, keeping it near the outer sides of the feeding head F, as shown at c.

In feeding heavy castings, there should be several feeding rods on hand, a few of them smaller than the ones generally used, so that if one rod becomes badly clogged, it can be

FIG. 13.

removed and a clean one used. Feeding rods may often be kept from becoming clogged by tapping them lightly with another rod held in the hand, as shown in Fig. 13. This jars off the metal that clogs the rod, and assists in keeping the feeding head open until the casting has solidified. The partially solidified iron that is jarred off the feeding rod can be worked into the casting; this takes it out of the way of the feeding and still benefits the casting by helping to cool the metal.

17. Hot Iron for Feeding.—In using feeding rods, care should be taken to have plenty of hot metal in order to keep the feeder open, so that the metal in the feeding head may be as fluid as that in the casting below the head. There is altogether too much disregard of this important point, and one frequently sees molders in difficulty from the start for want of hot iron to keep the feeding heads open. Where one has hot metal when necessary, feeding should be continued until the gradual solidification of the casting from the bottom upwards has pressed the feeding rod up to the lower edge of the feeding head. This stage reached, hot metal should be poured into the feeder while the feeding rod is being gradually removed. Any one feeding in this manner will obtain a solid casting.

Fig. 13 shows a molder in the act of feeding a mold and a second person pouring hot metal out of a hand ladle to keep the feeder open and supply the shrinkage of metal in the casting. With very large feeding heads, it is sometimes a good plan to use a small iron scoop and dip out the dull metal in the head for a depth of about 6 inches. The feeding head is then filled with hot metal as direct from the cupola as possible. This can be churned up and down with the feeding rod to open the feeding head, and so obtain a good clean hole to continue the feeding. If in feeding a casting the molder will keep his feeding rod hot and have a covering of powdered charcoal over the feeding head, he will experience little or no difficulty from iron adhering to the feeding rod. To be able to keep a feeding head open until the solidifying metal drives the rod out of the casting is an operation requiring skill on the part of the molder.

18. Use of Large Feeders.—Often with feeders over 8 inches in diameter, one will not need to insert a feeding rod for quite a time after the mold has been poured, for the reason that the iron does not commence to shrink until it is approaching the point of solidification. This is not until the metal in contact with the walls of the mold or core has cooled down considerably.

Many molders when feeding large heads like that in Fig. 10 cover the metal in the feeding head with dry blackening or charcoal dust as soon as the mold is poured, to keep the heat from escaping, and then pour in hot metal occasionally, as the feeding head settles. This may be continued in large feeders of over 12 inches in diameter for from 30 to 60 minutes, or as long as it is safe to do so. This can be determined by occasionally passing a hot feeding rod into the casting and observing whether any metal sticks to it when pulled out. When iron commences to stick it is time for the molder to put in the feeding rod for constant and careful operation, or until it is driven upwards and out of the casting by the solidifying metal, as already described.

BENCH MOLDING.

APPLIANCES AND PROCESSES.

19. Advantages of Bench Molding.—In order to save time and labor in making some kinds of small castings, benches are used to support the flasks during molding. There are many firms that make a specialty of bench molding with snap flasks only. The workmen are called **bench molders.** Some bench molders will make from 75 to 100 molds per day, according to the size of the flask and the character of the pattern used. The chief skill involved in bench molding generally lies in getting up the pattern and mold board. In some cases, a dozen or more patterns may be attached to a plate or pattern board, and the whole finished in such a manner that as soon as the plate is withdrawn from the mold, the cope can be put in place. While there are many founders making a specialty of bench molding only, there are many large foundries that could utilize benches and snap flasks for making some of their very light castings, instead of molding them on the floor after the manner of heavy work. Not only are benches used for convenience in ramming up snap flasks, but also for the ordinary small wooden and iron flasks, ranging from 12 to 14 inches, square or round. Snap flasks, however, are used very largely in bench molding and properly belong to this class of foundry work.

20. Snap Flasks.—A very large number and variety of small light castings are made in what are called **snap flasks.** For common use on the bench, these flasks range

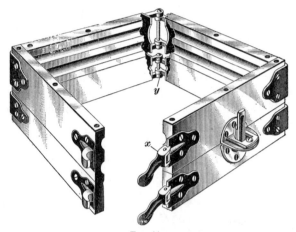

FIG. 14.

from 10 to 16 inches square, if of the form shown in Fig. 14, and from 10 to 16 inches in diameter, if of the form shown in Fig. 15. For ramming molds on the floor, larger sizes are used. Snap flasks may also be oblong or of any other shape desired. Those shown in Figs. 14 and 15 are constructed with clasps and hinges, as seen at x and y, so that they can be opened and removed from the outside of the mold without breaking it. This

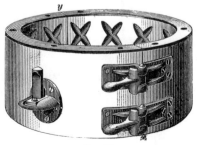

FIG. 15.

permits the use of one flask to make any number of molds.

It is chiefly molds that have little side pressure when being poured that are cast in snap flasks; for when the flask is removed from the mold there is little to prevent the pressure of the fluid metal from bursting the mold outwardly. To prevent snap-flask molds bursting, the molder places cribs or frames around the mold, which keep the sand in place. Snap flasks have no cross-bars, and hence cannot be used for large flat work. Also, it is impossible to use snap flasks for any piece that has a large lifting area. For instance, a stove door should not be cast in a snap flask on account of the large area over which the iron can exert a lifting influence tending to separate the cope and drag. Large flat pieces that have a considerable portion of their area cut away by holes, such as stove-door frames, etc., can be cast in snap flasks on account of the small lifting area which they present to the cope.

21. Bench Rammers.—In ramming flasks on the bench, two wooden rammers, shown in Fig. 16, are gener-

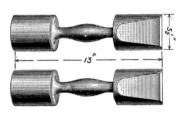

FIG. 16.

ally used, one in each hand. In ramming such flasks, the drag especially should be peened very solidly around the outer part of the mold, so as to give as much strength as practicable, to withstand the side pressure of the metal on the mold when it is poured.

In using a rammer in each hand, one should be as effective as the other; this requires practice.

22. Making a Bench Mold.—Fig. 17 (*a*), (*b*), (*c*), and (*d*) shows the first operations of bench molding as carried on in a stove foundry. The drag and cope of a snap flask are shown at *a* and *d*, Fig. 17 (*a*). The pattern *b* is placed on the molding board *c*, ready for use. The drag *d* is placed on the molding board with the dowel-pin *e* of the drag between the nails, or other fastening, on the molding

FIG. 17.

board, as shown at *f*, Fig. 17 (*a*). The molder sets his

FIG. 18.

riddle *g* on top of the flask and shovels it full of sand from

FIG. 19.

the pile under his bench. He then picks up the riddle and riddles the sand over the pattern, as shown in Fig. 17 (*b*), until there is an inch or so over the face of the pattern, after which he dumps the rest of the sand in the riddle into the flask and shovels in sand until there is quite a heap above the flask. He then rams all around the edge of the mold close to the flask with the handle of the

Fig. 20.

shovel k, as shown in Fig. 17 (c).   All the motions are quick
and forcible.   Next he takes the bench rammers l and rams
the sand all over the middle of the mold with these rammers
held flat, pushing both rammers down together, as shown
in Fig. 17 (d).   As soon as he has gone over the mold once or
twice in this way, he rams down hard with the butts all over
the mold, as shown in Fig. 18.   The next operation is to
strike off the mold, which is done by taking a straightedge
and sawing it back and forth across the top of the mold in a
zigzag manner, leaving the top smooth.   The top is then

Fig. 21.

sprinkled over with a handful of molding sand, and the bottom board is placed on and rubbed down with a rotating
movement to a good smooth bed on the sand and against
the flask.   The flask is then grasped firmly on both sides,
drawn to the edge of the bench, and as it drops off is rolled
over with a quick flop on to the bench, as shown in Fig. 19.
The molding board is then removed, the top of the mold
cleaned off and smoothed up, parting sand sprinkled over the
mold, the cope a fitted on to the drag d, and the sprue pin m
put in place, as shown in Fig. 20 (a).   Sand is again riddled

all over the mold and rammed as before.   The top is
smoothed off with the strike *o* and the sprue pin *m* removed
with a draw-spike *n*, as shown in Fig. 20 (*b*).

(*b*)

(*a*)

FIG. 22.

After the sprue pin is withdrawn, the sand around the top
of the sprue is rounded off with the fingers and packed firmly.
The cope is then lifted off carefully, so that the sand shall

not be loosened, and placed standing on edge on the bench, as shown at *a*, Fig. 20 (*c*). The bellows *h* is used to blow off any loose sand or dust around the pattern or on top of

FIG. 23.

the drag, as shown in Fig. 20 (*c*). With the sponge and quill *p*, water is dripped all around the sand near the edge of the pattern, as shown in Fig. 20 (*d*). The sponge is fastened to the quill, so that when the sponge is full of water it can be squeezed in such a way as to let a few drops off the end of the quill or to let quite a little stream of water from the sponge. This moisture on the sand makes it firmer, so that the pattern can be removed without displacing the sand. The pattern is now ready to draw, which is done very carefully with draw-pins, as shown in Fig. 21. The pattern for the gates is then removed, as gates are cut to connect the sprue with the mold, and the mold is smoothed with a finishing tool if there are any defects. The cope is then replaced, the snap flask removed by loosening the catches *q* and *r*, and both cope and drag are taken off at the same time, as shown in Fig. 22 (*a*). The flask is placed on its side in a convenient position, and a crib *s*, Fig. 22 (*b*), put on the mold to keep it in good shape while pouring. The mold is picked up, carried out on the floor, and placed in the row with the molds already there, as shown in Fig. 23. When the day's molding has been finished the molds are poured, as shown in Fig. 24. This illustration gives a good idea of how a foundry floor looks when the molds are being poured.

FIG. 24.

23. Pouring Off. — In foundries where small work forms the largest part of the output, it is customary to stop the molding in the early part of the afternoon and pour off the molds that have been made since morning. The molten metal is brought to the molder usually in traveling ladles, but sometimes by hand ladles. Any dirt that can be skimmed off is removed, and the molder pours the molten metal into the mold, as shown in Fig. 24. A cast-iron weight t is placed on the mold before pouring, and the molder holds the ladle as low as possible and takes care not to strike it against anything. He brings the ladle as close to the mold as he can without touching it, and turns the ladle gradually until a small stream of metal starts into the mold. By watching the metal closely, he is able to stop pouring before the metal runs over the side of the mold. When he has poured all the molds that he can pour completely with the one ladleful, the balance is poured into small pig molds, which are merely small cast-iron **V** troughs about 10 inches long. The weights t, t are then shifted to other molds before pouring the next ladle of metal. The purpose of these weights is to prevent the metal from lifting or displacing the copes on the molds. The letters in Fig. 24 have the same signification as in the previous figures.

24. Shaking Out and Tempering the Sand. — The pouring having been finished, the molder starts in to remove the castings from the sand. This process is commonly called shaking out the castings. The cribs s, s are removed from the mold and placed in a pile w. The casting is removed with tongs and placed to one side in a pile, and the sand is cleaned off the bottom board u, which is placed on the pile of bottom boards v. When all the castings have been removed and the sand cleaned off the bottom boards, the sand is tempered as previously explained, and piled in the middle of the floor. The bench on which the molder has his tools is moved over the sand pile, so that when commencing work the next day the molder has simply to turn around and put his first mold in place on the floor. As he works he shovels the sand from under the bench, and the bench is moved along so that he can keep up with the sand as he uses it and at the same time have room for the molds as they are finished. By working in this manner the molder does not have to go far for his sand nor to place the molds. This is a great convenience and it saves the molder time. In some cases a wheel is placed on each of the back legs, so that the molder can move the bench back without the help of another man.

The arrangement of tools about the bench and on it is also for the purpose of saving time. The cope a, the drag d, the molding board c, the riddle g, the shovel k, the hand rammers l, the sprue pins m, the draw-spike n, the strike o, the cribs s, the bottom boards v, a brush y for clearing off dirt, and a tool box z, in which small finishing and other tools are kept, are all on or near the bench. In this way everything is convenient and handy for rapid and expeditious work. For some classes of bench molding, the benches are stationary, being placed along the wall near the windows. In such cases it is necessary to bring the sand to the bench and take the molds away.

PROTECTING THE MOLD AGAINST FUSING.

25. Composition of Sand and of Protective Material.—In order to obtain a smooth surface on the casting, it is essential that the sand forming the face of the mold should be well tempered and sieved. For a large class of work the sand should be mixed with some protective material to keep the sand from fusing. Molding sands are composed of silica and alumina, with small quantities of lime, oxide of iron, potassium, and magnesia. The three latter elements are easily fused; they combine with the silica to form silicates or a kind of glass that on heavy castings may form a scale varying in thickness from $\frac{1}{100}$ inch up to $\frac{1}{2}$ inch, if the face of the mold is not protected with some non-fusible material. The face of the mold may be protected by mixing finely ground **sea coal** with the sand, or by facing the mold with blackening. In some cases a facing sand containing sea coal is used, and this surface is covered with blackening.

Sea coal in America outside the Pennsylvania coal district is obtained chiefly from the culm, or slack, of bituminous coal. In addition to mixing sea coal with the facing sand, the surfaces of molds are often covered with what is called **blackening** or **lead.** Blackenings and leads consist chiefly of carbon, the other ingredients being alumina, silica, lime, and iron. The less of these latter elements present, the more intense heat the blackenings and leads will stand before fusing. The cheaper blackenings are composed chiefly of coal dust, or culm, with the addition of various minerals. These blackenings when ground to a powder lack cohesion, and, therefore, they are apt to float or wash before the iron when the mold is being poured. To guard against this, various minerals or clays are ground in the blackenings to give them cohesion. The finer the pure coke or carbon is ground, the more adhesion will it possess; and when ground very fine, the adhesion may be sufficient to hold the material together without the use of any bond. Lehigh blackenings are made by grinding to a dust fine qualities of Lehigh coal. Coke blackenings are made from good grades of coke, selected from a grade having the highest fixed carbon, which at times runs as high as 90 per cent. Gas-house carbon being practically a pure carbon, makes excellent blackening; but, on account of its being a difficult material to grind and bolt, it is not in much favor with blackening manufacturers. In order that charcoal may make a good blackening, it must be made from some hard wood, as hard maple, and be carefully burned. Soft or stringy-grained wood is useless for making charcoal blackenings.

Plumbago or **graphite** is recognized as being the most non-fusible material used for blackening. The best grade of imported graphite comes from the island of Ceylon. In its crude state it looks like bright chips of burnished silver, from which fact it is commonly called *silver lead.* This grade of graphite is not only very valuable for mixing with blackenings intended for green-sand, dry-sand, and loam work, but it is also good to dust in a dry state over the surface of molds, as it assists in sleeking the mold and peeling the casting. Of recent years, graphite has been extensively used in many foundries. In putting it on a mold, many molders use camel's-hair brushes, while others will shake it out of a bag or throw it on by hand. High grades of graphite excel all other blackening in peeling a casting and giving it a fine, smooth face, and bright color. Some grades of graphite are procurable in this country. North Carolina produces some graphite, but it is mixed largely with clay and other foreign substances. Graphite is also found in eastern Pennsylvania and Tennessee, but the best grade comes from Ticonderoga, New York.

Blackening materials are sometimes called leads, on account of the fact that graphite is called *black lead* or *silver lead.* Blackening materials never contain metallic lead.

PREPARING THE FACING SAND.

26. Percentage of Sea Coal in Facing Sand.— As a rule, the heavier the casting, the more sea coal is required in the facing, the limit being 1 part of sea coal to 6 parts of sand. If more sea coal than this is used on any casting, it is very liable to make the surface streaked or veined, especially in heavy work. Where the molds are heavily coated with good graphite, or are poured with dull iron, this will not be so pronounced. If facing sand contains too much sea coal, it will prevent light castings running sharply, and is very apt to cause **cold shuts.** This is the condition induced when two bodies of fluid metal run together but fail to unite. The surface of the casting will be harder than if a weaker facing had been used; this is due to the gas that sea coal generates when the mold is being poured. This gas is liable to create a cushion between the mold and the metal, and the amount of cushioning depends on the speed of pouring, the amount of pressure, and the fluidity of the metal. Where the facing sands contain too much sea coal and the castings are poured *dull*, the metal often becomes set before the gas cushions are all destroyed by the gas passing out through the vents. This prevents the metal from running into the corners and edges of the mold and may also cause cold shuts. Then again, these gases may make smooth concave indentations in castings, as seen at *w*, Fig. 7. Another effect sometimes produced by sea coal is the coating of the surface of the castings with what might be termed coal soot. However, in order that this shall take place to any great extent, a combination of conditions that seldom occurs must take place. While the objectionable conditions described are usually to be found in light-weight and medium-weight castings, still, heavy castings, when poured with dull iron, may also present some of them.

The proportion in which sea coal is mixed with sand ranges from 1 in 20 to 1 in 6. Castings under $\frac{3}{4}$ inch in thickness seldom require any facing sand. Below $\frac{3}{4}$ inch thickness, better and smoother castings are often obtained by using common heap sand, well tempered and sieved finely on the patterns, especially when the metal cannot be poured quickly. In general, castings from $\frac{3}{4}$ inch to 1 inch thick require facing sand having 1 part of sea coal to 12 parts of sand; above 1 inch and up to 2 inches, 1 part sea coal to 10 parts sand; from 2 inches to 4 inches, 1 part sea coal to 8 parts sand; all above 3 inches in thickness, 1 part sea coal to 6 parts sand. In mixing facing sand, some molders use common heap sand, or old sand mixed with new in varying proportions, according to the strength of the different sands.

It is not always the thickness of the casting that regulates the strength of the facing sand. There are many other things to be considered: (1) whether the casting is to be poured with hot or dull iron; (2) the distance of some parts of the mold from the point where the metal enters; (3) the

time it will take the mold to become filled with iron;
(4) whether the metal is running over flat surfaces, and
(5) is covering them slowly or quickly.   Strong facings on
the sides of a mold, where the iron runs in and rises slowly,
may cause heavy castings to be cold shut.   The square
corners of castings should, generally speaking, have weaker
facings on them than the straight plain surfaces, and the
lower parts of high molds should have a stronger facing
than the upper portion.   If the strong facing suitable for the
lower portion were used at the upper portion, the casting at
the upper part would be curly or partly cold shut at the surface, owing to the dulness of the metal when it reached the
upper portion of the mold.   A new sand without mixture
will require more sea coal than if it were mixed with old
or common heap sand.

Where molds are long in preparation, or are to withstand
rough usage in drawing the patterns, it is best, if there are
no conditions prohibiting it, to use new sand in mixing the
facings.   For copes covering heavy castings, it is best to use
all new sand.   For light-work copes that cover flat surfaces
it is best, when the new sand is strong, to use some old sand
with it, sometimes using equal proportions.

**27. Necessity of Thoroughly Mixing Facing
Material.**—It is not uncommon to find castings streaked,
veined, cold shut, or not peeled, because the sea coal in the
facing had not been thoroughly mixed with the sand.   When
castings are very massive, as great a percentage of sea coal
as possible is mixed with the sand, in order to peel the
casting.   Much more sea coal can be added to some sands
by thorough mixing than would otherwise be possible.   In
some cases it may be possible to mix facings for massive
castings with as high a proportion as 1 part of sea coal to
5 parts of sand, provided the sea coal is thoroughly mixed
with the sand.   A good way to mix facing sands thoroughly
is to tramp them with the feet for from 2 to 4 cuttings or
turnings over of the sand, and then at the last cutting to
riddle it through a $\frac{1}{2}$-inch riddle and then through a $\frac{1}{4}$-inch

riddle, and finally through the sieve before applying it to the face of the pattern. In cutting the sand, it should be well scattered, as shown in Figs. 1 and 2, *Green-Sand Molding*, Part 1. The best way to mix sand, however, is with a sand grinder, which is a machine for breaking up lumps and mixing the sand thoroughly.